BIOCHEMISTRY RESEARCH TRENDS

GLYCINE

BIOSYNTHESIS, PHYSIOLOGICAL FUNCTIONS AND COMMERCIAL USES

BIOCHEMISTRY RESEARCH TRENDS

Additional books in this series can be found on Nova's website under the Series tab.

Additional E-books in this series can be found on Nova's website under the E-book tab.

CHEMICAL ENGINEERING METHODS AND TECHNOLOGY

Additional books in this series can be found on Nova's website under the Series tab.

Additional E-books in this series can be found on Nova's website under the E-book tab.

BIOCHEMISTRY RESEARCH TRENDS

GLYCINE

BIOSYNTHESIS, PHYSIOLOGICAL FUNCTIONS AND COMMERCIAL USES

WILHELM VOJAK
EDITOR

New York

For permission to use material from this book please contact us:
Telephone 631-231-7269; Fax 631-231-8175
Web Site: http://www.novapublishers.com

Library of Congress Cataloging-in-Publication Data

Library of Congress Control Number: 2012952221

ISBN: 978-1-62417-263-2

Published by Nova Science Publishers, Inc. † New York

Contents

PREFACE

In this book, the authors present current research in the study of the biosynthesis, physiological functions and commercial uses of glycine. Topics discussed include the molecular complexation of glycine with 18-crown-6 in aqueous-organic solvent; concentrated glycine based electrolytes and application to solid pH sensor; heterologous expression of the glycine receptor chloride channel; the theoretical evaluation of the properties of energy deposition associated with the collision of fast ions with glycine and its zwitterions; electrochemistry of aliphatic bifunctional molecules; and the clinical uses of glycine.

Chapter 1 – In this chapter the authors discuss the influence of water-ethanol (H_2O-EtOH), water-dimethylsulfoxide (H_2O-DMSO) and water–acetone (H_2O-MeAc) mixed solvents on the molecular complex formation reaction between biological model substances glycine (Gly) and 18-crown-6 (18C6), 18C6+Gly ↔ [Gly18C6]. The analysis of the interdependence between the thermodynamic reaction parameters (Gibbs energy, enthalpy and entropy) of complex formation and the thermodynamic parameters for the solvation of reagents allows to reveal regularities and peculiarities of the influence of solvent composition on formation of [Gly18C6].

An increase of lgK^{o}([Gly18C6]) and of exothermicity for $\Delta_r H^{\circ}$([Gly18C6]) with increasing the non-aqueous component mole fraction in mixed solvents is observed as a common effect for all studied media. The discussion and interpretation of this phenomenon is based on the solvation-thermodynamic approach.

The analysis of the solvation enthalpy contributions of all reaction participants, $\Delta_{tr}H^{\circ}$(Gly), $\Delta_{tr}H^{\circ}$(18C6), and $\Delta_{tr}H^{\circ}$([Gly18C6]), to the reaction enthalpy transfer, $\Delta_{tr}H^{\circ}_r$, shows that the strengthening of the desolvation of

18C6 and Gly with increasing the concentration of EtOH, DMSO or MeAc in the mixed solvents favors the increase of the reaction exothermicity. The thermodynamic ratio α_{dif} determined by means of the solvation-thermodynamic approach is used for estimating the enthalpy changes for the complexation reaction in the studied solvents according to: $\Delta_{tr}H^{\circ}_{r,\ calc} = (\alpha_{dif} - 1)\Delta_{tr}H^{\circ}(18C6)$.

The good agreement between $\Delta_{tr}H^{\circ}_{r.calc}$ and $\Delta_{tr}H^{\circ}_{r}$ in the examined composition range of the mixed solvents allows to estimate the influence of H_2O-EtOH, H_2O-DMSO, H_2O-MeAc solvent composition on the energetics of complex formation at the solvent compositions with high concentration of non-aqueous components where the very low solubility of glycine makes impossible the experimental determination of the reaction enthalpy.

The relationship between the solvation contributions of the reagents for the transfer of reaction Gibbs energy from water into H_2O-EtOH and H_2O-DMSO solvents is essentially different from the corresponding values of the reaction enthalpy. The analysis of the solvation contributions of reagents into the $\Delta_{tr}G^{o}_{r}$ in H_2O-EtOH and H_2O-DMSO solvents reveals the prevailing role of the desolvation contribution of Gly, $\Delta_{tr}G^{o}$(Gly), in determining the Gibbs energy of transfer of reaction from water in water-organic solvents.

The behavior of the thermodynamic parameters of molecular complex formation reaction with 18C6 of glycine is compared with that of D,L-alanine and L-phenylalanine as well as with thermodynamic data of complexation between 18C6 and silver (I) ion in non aqueous media.

Chapter 2 – The anodic oxidation of concentrated based glycine electrolyte on smooth platinum electrode yields a strongly grafted polyglycine-like coating. This electrochemical reaction occurs in an irreversible way on platinum but also on other material such as graphite. Contrary to studies performed in diluted medium, spectroscopic surface analysis revealed the presence of peptide bonds in the resulting thin film layer on platinum with polyglycine like (type II) structure. Since the electrodeposited polyglycine like coating at alkaline pH have chemical groups with affinity towards protons, it was used as transducer and embedded in a solid potentiometric pH sensor. Local pH measurement can be achieved by using miniaturized electrodes thanks to photolithography process. The couple (silver chloride as reference electrode, polyglycine based platinum electrode as working electrode) of microelectrodes gives linear potentiometric responses versus pH in the range [2 - 12], reversibly and with a sensitivity of 52.4 mV/pH (for 1 mm electrode diameter). This solid pH sensor was tested over a period of thirty days and its potentiometric behavior was compared to other polymer based pH electrodes such as linear polyethylenimine (L-PEI) and polyaniline (PANI). Quality control tests can be considered by combining miniaturized pH sensors with other measurements.

Chapter 3 – The glycine receptor (GlyR) is a neuronal chloride channel, which, upon binding glycine, allows chloride anions to permeate into the neuron causing hyperpolarization of the transmembrane potential and inhibition of neuronal activity. To investigate the structure and function of the GlyR, scientists usually isolate it by expressing it in a heterologous expression system, such as the *Xenopus laevis* oocyte or a cultured immortalized mammalian cell line, such as HEK 293. In this chapter, protocols for the expression of the GlyR in these two expression systems are given, including preparation of recombinant cDNA, preparation of cRNA, culturing HEK 293 cells, surgical isolation of *Xenopus laevis* oocytes, transfection of cDNA into the HEK 293 cells, injection of cRNA into the oocytes, and detection of the expressed GlyRs.

Chapter 4 – With the advent of the use of precise ion accelerators for medical purposes, it becomes ever more important to understand the interaction of biomolecules with fast ions. Glycine is both a protein component and a model biomolecule, and is thus an important test system. In this report, the authors discuss various aspects of the theoretical evaluation of the properties of energy deposition associated with the collision of fast ions with glycine and its zwitterions, including differences due to molecular conformation and orientation with respect to the ion beam direction as well as due to the effect of surrounding water molecules and the state of target aggregation. Quantum mechanical calculations, which yield the dipole oscillator strength distribution of glycine are reported. The ease with which energy is absorbed from a fast ion, described by the mean excitation energy and stopping power of glycine, is reported and discussed.

Chapter 5 – Despite their toxicity and the rise of green chemistry, nonaqueous solvents play an important role in analytical electrochemistry. When used methodically, they can bring an elegant way to modify a substrate irreversibly in one-pot experiment. Electrochemistry in liquid ammonia and liquid amines contributes in this direction too. Since ammoniates and aminates (by analogy to the name hydrate for aqueous electrolytes) are very used in the authors' group, they are doing electrochemistry in these media:

1. to elucidate new (electrochemical) reactions, electropolymerization, etc.

2. to better discriminate the analytes by use of several techniques in (bio) chemical sensors (bio) analytical
3. to model and interpret the results by theoretical ab initio and docking.

The authors will explain, in what follows, the three points that summarize their research the authors initiated and developed in Besancon, France.

Chapter 6 – Glycine is a non-essential amino acid used commonly asa constituent of amino-acid based nutritional supplements (enteral formulas) in medical settings. As an example, standard infant nutritional formulas all contain glycine (1-2 g/100 g protein), in an effort to match breast milk - which contains 2.7 g glycine per 100 g protein (http://www.brightbeginnings.com/ -- Feb 2007), which amounts to approximately 35 to 40 mg per 5 floz feeding. Glycine is alsoused to lavage the genitourinary tract during gynecologic and urologic surgery. More recently, in the treatment of pulmonary hypertension, glycine has been used as an alkalinizingagent for inhaled prostacyclin, which requires alkaline conditions for stability. Presently, studies are evaluating the use of inhaled isotonic alkaline glycine buffer as a diluent for many inhaled therapeutic compounds and the use of intravenous glycine for amelioration of oxidative stress and ischemia/reperfusion injury.

In: Glycine
Editor: Wilhelm Vojak
ISBN: 978-1-62417-263-2

Chapter 1

Molecular Complexation of Glycine by 18-Crown-6 in Aqueous – Organic Solvents: A Solvation -Thermodynamic Study

T. R. Usacheva[1*], V. A. Sharnin[1] and E. Matteoli[2,#]
[1]Ivanovo State University of Chemistry and Technology, Ivanovo, Research Institute of Thermodynamics and Kinetics of Chemical Processes, Russia
[2]CNR-IPCF, Institute of Physical-Chemical Processes of National Research Council, Pisa, Italy

Abstract

In this chapter we discuss the influence of water-ethanol (H_2O-EtOH), water-dimethylsulfoxide (H_2O-DMSO) and water–acetone (H_2O-MeAc) mixed solvents on the molecular complex formation reaction between biological model substances glycine (Gly) and 18-crown-6 (18C6), 18C6+Gly ↔ [Gly18C6]. The analysis of the interdependence between the thermodynamic reaction parameters (Gibbs energy, enthalpy and entropy) of complex formation and the thermodynamic parameters

* Email:oxt@isuct.ru.
# Email:matteoli@ipcf.cnr.it.

for the solvation of reagents allows to reveal regularities and peculiarities of the influence of solvent composition on formation of [Gly18C6].

An increase of *lgK*°([Gly18C6]) and of exothermicity for $\Delta_r H°$([Gly18C6]) with increasing the non-aqueous component mole fraction in mixed solvents is observed as a common effect for all studied media. The discussion and interpretation of this phenomenon is based on the solvation-thermodynamic approach.

The analysis of the solvation enthalpy contributions of all reaction participants, $\Delta_{tr}H°$(Gly), $\Delta_{tr}H°$(18C6), and $\Delta_{tr}H°$([Gly18C6]), to the reaction enthalpy transfer, $\Delta_{tr}H°_r$, shows that the strengthening of the desolvation of 18C6 and Gly with increasing the concentration of EtOH, DMSO or MeAc in the mixed solvents favors the increase of the reaction exothermicity. The thermodynamic ratio α_{dif} determined by means of the solvation-thermodynamic approach is used for estimating the enthalpy changes for the complexation reaction in the studied solvents according to: $\Delta_{tr}H°_{r,\ calc} = (\alpha_{dif} - 1)\Delta_{tr}H°(18C6)$. The good agreement between $\Delta_{tr}H°_{r.calc}$ and $\Delta_{tr}H°_r$ in the examined composition range of the mixed solvents allows to estimate the influence of H_2O-EtOH, H_2O-DMSO, H_2O-MeAc solvent composition on the energetics of complex formation at the solvent compositions with high concentration of non-aqueous components where the very low solubility of glycine makes impossible the experimental determination of the reaction enthalpy. The relationship between the solvation contributions of the reagents for the transfer of reaction Gibbs energy from water into H_2O-EtOH and H_2O-DMSO solvents is essentially different from the corresponding values of the reaction enthalpy. The analysis of the solvation contributions of reagents into the $\Delta_{tr}G°_r$ in H_2O-EtOH and H_2O-DMSO solvents reveals the prevailing role of the desolvation contribution of Gly, $\Delta_{tr}G°$(Gly), in determining the Gibbs energy of transfer of reaction from water in water-organic solvents. The behavior of the thermodynamic parameters of molecular complex formation reaction with 18C6 of glycine is compared with that of D,L-alanine and L-phenylalanine as well as with thermodynamic data of complexation between 18C6 and silver (I) ion in non aqueous media.

INTRODUCTION

Problems of molecular complex formation in systems containing solvated biomolecules and macrocyclic ligands are of current interest for fundamental studies and their solution is of considerable importance for applications. Molecular complex formation reactions occur in many biochemical processes such as enzymatic catalysis, membrane transport, and antigen-antibody interactions, which take place in aqueous solutions [1-5]. Molecular complexes consist of two or more molecules or ions that are bonded together in a unique structural system mainly due to electrostatic interactions such as hydrogen bonds, ion pairs, π-interactions, van der Waals forces; phenomena like the restructuring of the solvent, preferential solvation and partially formed or broken covalent bonds (transition state) may also been involved. The high structural organization is usually implemented only through multiple binding sites. It is known that the selection of "guest" molecules by "host" macromolecules to form of the molecular complex compound in aqueous solutions is determined by the complementarity principle. At the same time, reactivity of reagents, reaction energy characteristics and rates depend on the solvation state of chemical reaction participants in solution [6]. A series of studies of complex formation between macrocycles and amino acids in different media has shown these dependences [7-22].

The complex formation in water is often characterized by the high hydration properties of water as medium of the processes. As a consequence, low stability of molecular complexes in water is expected [7-9], which limits the synthesis and applications of new coordination compounds based on macrocycles. On the contrary, the molecular complex formation reactions in pure non-aqueous solvents show high complex stability and strong reaction exothermicity [10-22]. Non-aqueous media can promote the formation of molecular complexes in solution, helping to provide insight on the nature of the "host - guest" interaction in biological reactions. Crown ethers are efficient macrocyclic ligands ("host") capable of selectively coupling metal ions and neutral molecules, incorporating them into the internal cavity of its ring-like structure (figure 1). The first works devoted to the synthesis and the complexing properties of crown ethers have been published in the 1967-1970 years by Pedersen [23, 24]. Crown ethers are macrocyclic polyethers having atoms of oxygen as electron donors in its monomer units (-CH_2-CH_2-O-). In some cases, some of the O atoms are replaced by N or S. The shape of these molecules resembles a crown, which determined their name.

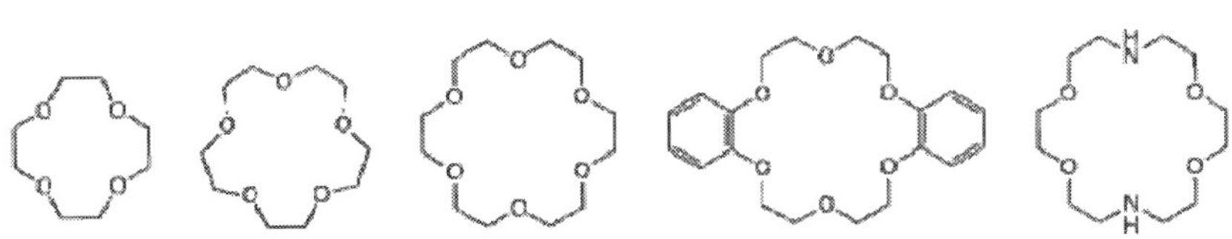

12-crown-4 15-crown-5 18-crown-6 dibenzo-18-crown-6 diaza-18-crown-6.

Figure 1. Structures and denomination of common crown ethers.

The oxygen atoms form a hydrophilic cavity, and the methylene hydrogen atoms form a hydrophobic outer shell. The use of crown-ethers and other macrocyclic compounds as a class of complexing agents for metal ions demonstrated to be a powerful means for the understanding of coordination chemistry of reactions in biological systems as well as of organic synthesis and catalysis [25-45].

The interactions of cations with macrocyclic ligands are similar to the process of solvation of ions by the molecules of polar solvents. Macrocyclic ligands act as a structured solvation shell for cations. All donor atoms of the macrocycle can be involved in the complex formation; this is not always possible in the coordination of cations by non-cyclic ligands with the consequence of loss in the stability of the complexes.

Glycine (Gly) is the simplest aliphatic amino acids, the only one of all protein constituents which has no asymmetric carbon atom. The principal biological function of glycine is as a precursor to proteins. It is also a building block to numerous natural products. At physiological values of pH (5.5 ÷ 7.0) amino acids do not carry a neat charge being bipolar (zwitterionic form). In particularly, for Gly the equilibrium between neutral and zwitterionic form is

$$NH_2CH_2COOH \leftrightarrow {}^{+}NH_3CH_2COO^{-} \quad (1)$$

(neutral form) (zwitterionic form)

In this chapter we analyze the influence of water-ethanol (H_2O-EtOH), water- dimethylsulfoxide (H_2O-DMSO) and water-acetone (H_2O-MeAc) mixed solvent composition on the thermodynamic parameters (ΔG, ΔH, ΔS) of molecular complex formation between 18-crown-6 ether (18C6) and glycine:

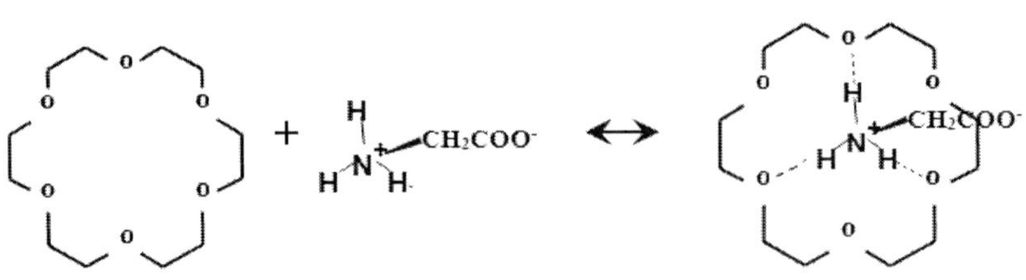

$$18C6_{solv} + Gly_{solv} \leftrightarrow [Gly18C6]_{solv} \quad (2)$$

The thermodynamic data of reaction (2) obtained in H_2O-EtOH, H_2O-DMSO and H_2O-MeAc solvents are compared with those of complex formation reaction of 18C6 with D,L-alanine (Ala) and L-phenylalanine (Phe) in aqueous-organic solvents as well as with thermodynamic data of complexation between 18C6 and silver (I) ion in non aqueous media. An analysis of the influence of changes in the solvation state of 18C6, amino acids and molecular complexes on the thermodynamic parameters of the corresponding complex formation reaction is also carried out. To this purpose, the solvation-thermodynamic approach [6] is used.

THE APPLICATION OF THE SOLVATION-THERMODYNAMIC APPROACH TO THE ANALYSIS OF MIXED SOLVENT INFLUENCE ON THE [GLY18C6] COMPLEX FORMATION

That solvent influences solvation and complex formation equilibrium of ions and individual molecules is a well known fact. When the solution medium is a mixture of two or more solvents, the phenomenon of preferential solvation may take place, in which the solvation shell of the solute has a composition different from that of the bulk solvent mixture, and model for its calculation have been proposed [6, 47-49]. The effects of mixed solvent composition and of the preferential solvation on the change in the solvation state of metal ions have been studied in detail [50-54]. When a reaction happens in a mixed solvent, each participant may have its own preferential solvation which can also depend on solvent composition. In this way, the reactivity of each participant strongly depends on the chemical nature and the composition of mixed solvents.

The estimation of the influence of mixed solvent composition on the thermodynamic solvation parameters of reagents and products (ΔG, ΔH, ΔS), and the derivation of proper relationships between their solvation parameters and those of reaction, allow to devise predictive models for complex formation and therefore to select the proper solvent for the synthesis of molecular complexes with predetermined thermodynamic properties such as stability and energetic of complex formation reaction.

Studies of the formation of amine, carboxylate and crown-ethers complexes with *d*-metal ions in aqueous-organic solvents have revealed, on the base of the solvation-thermodynamic approach, several general rules governing the thermodynamic characteristics of reactions and solvation of the reagents [46, 55-60]. These rules have shown the possibility of predicting the thermodynamic parameters of the ionic complex formation reactions in different media according to a change in the solvation state of ligands.

According to the solvation-thermodynamic approach, the transfer of the reaction (2) from water to a mixed solvent, (s), can be calculated referring to the following thermodynamic cycle:

$$\begin{array}{ccccc} & & \Delta_r Y^\circ_{(H2O)} & & \\ Gly_{(H2O)} & + \; 18C6_{(H2O)} & \rightarrow & & [Gly18C6]_{(H2O)} \\ \downarrow\Delta_{tr}Y^\circ(Gly) & \downarrow\Delta_{tr}Y^\circ(18C6) & \downarrow\Delta_{tr}Y^\circ_r & & \downarrow\Delta_{tr}Y^\circ([Gly18C6]) \\ & & \Delta_r Y^\circ_{(s)} & & \\ Gly_{(s)} & + \; 18C6_{(s)} & \rightarrow & & [Gly18C6]_{(s)} \end{array}$$

where $\Delta_{tr}Y^\circ_r$ and $\Delta_{tr}Y^\circ$ are the thermodynamic function of transfer ($\Delta_{tr}G^\circ$, $\Delta_{tr}H^\circ$, $\Delta_{tr}S^\circ$) for the reaction (2) and the reagents and product, respectively, from water to a mixed solvent (s).

These functions quantitatively reflect the influence of the change of solvent composition on the thermodynamic parameters and are calculated for the reaction:

$$\Delta_{tr}Y^\circ_r = \Delta_r Y^\circ_{(s)} - \Delta_r Y^\circ_{(H2O)} \qquad (3)$$

and for the participants (Z=Gly, 18C6, [Gly18C6]):

$$\Delta_{tr}Y^\circ(Z) = Y^\circ(Z)_{(s)} - Y^\circ(Z)_{(H2O)} \qquad (4)$$

where $\Delta_r Y^\circ_{(s)}$ and $\Delta_r Y^\circ_{(H2O)}$ are the thermodynamic parameters (ΔG°, ΔH^o, $T\Delta S^o$) of the complex formation (2) and $Y^\circ_{(s)}$ and $Y^\circ_{(H2O)}$ are the thermodynamic parameters of reagents solvation in a mixed solvent and in H_2O, respectively.

From the above thermodynamic cycle and eqs. (3) and (4), the solvent influence on the change of thermodynamic parameters of reaction (2) can be expressed in terms of the solvation contribution, $\Delta_{tr}Y^\circ$, of all participants, by the equation:

$$\Delta_{tr}Y^\circ_r = \Delta_{tr}Y^\circ([Gly18C6]) - \Delta_{tr}Y^\circ(Gly) - \Delta_{tr}Y^\circ(18C6) \qquad (5)$$

To apply eq. (5) for the reagents contribution analysis into the change of thermodynamic parameters of the reaction, the solvation thermodynamic parameters for reagents ($\Delta_{tr}Y°$(Gly), $\Delta_{tr}Y°$(18C6)) and the product, $\Delta_{tr}Y°$([Gly18C6]) must be known. Usually, the needed thermodynamic data of the reagents solvation can be obtained by different experimental methods, such as calorimetry (for enthalpic parameters), determination of reagent distribution between two non-mixed liquid phases, or solubility experiments (for the Gibbs energy property). If the solvation thermodynamic parameters of reagents and reaction are known, eq. (6) can be used to calculate the changes in the solvation state of the complex:

$$\Delta_{tr}Y°([\mathrm{Gly18C6}]) = \Delta_{tr}Y°_r + \Delta_{tr}Y°(\mathrm{Gly}) + \Delta_{tr}Y°(\mathrm{18C6}) \qquad (6)$$

Thus, it is not necessary to provide an additional experiment to obtain the thermodynamic parameters of the complex particle.

The analysis of the dependence of thermodynamic reaction functions, ($\Delta_{tr}Y°_r$), on the thermodynamic reagent solvation parameters, ($\Delta_{tr}Y°$), by means of eq. (5) allows to reveal the important role played by the solvation contribution, namely a key factor, in determining $\Delta_{tr}Y°_r$. After evaluating a key factor, it may be possible to trim the thermodynamics of chemical equilibrium by influencing the solvation state of the selected reagent through appropriate selection of solvent nature and composition.

An attempt to identify the key factors which drive the thermodynamics of the molecular complex formation between 18C6 and Gly has been done with the aim of devising solvation thermodynamic models capable to predict the thermodynamics of the reaction between 18C6 and amino acids in other solvents.

The Influence of H_2O-EtOH, H_2O-MeAc and H_2O-DMSO Mixed Solvent on the Stability of [Gly18C6]

The molecular complex formation (2) was studied by the calorimetric titration method at $T = 298.15$K.

The calorimetric experiments were carried out by means the microcalorimetric systems TAM Thermometric MOD 2277 (Sweden) in the Institute of Physical-Chemical Processes of National Research Council, CNR,

Pisa, Italy. They were performed in a large range of EtOH concentration (X_{EtOH} = 0.0 ÷ 0.91 mole fraction) and at a less broad range of concentrations of acetone (X_{MeAc} = 0.0 ÷ 0.21 mole fraction) and dimethylsulfoxide (X_{DMSO} = 0.0 ÷ 0.25 mole fraction). The solvent composition ranges of H_2O-MeAc and H_2O-DMSO are limited by low solubility of Gly at high concentration of MeAc and DMSO in aqueous mixtures. Almost all amino acids are fairly soluble in water, little soluble in water - organic solvent mixtures and insoluble in non-aqueous media [10, 61- 63]

The values of lgK^o([Gly18C6]) and Δ_rH^o([Gly18C6]) were simultaneously calculated by processing the calorimetric experimental data by a least square method.

The stoichiometric model of 1:1 complex formation satisfactorily described the experimental data and it is in agreement with the literature [8-11]. X-ray and potentiometric studies have shown that the amino acid molecule binds to the oxygen atoms of macrocyclic ligand with NH_3^+-group of zwitterionic amino acid form by means three hydrogen bonds [8-11] (see sketch of eq.(2)).

Table 1. Stability constant (lgK^o) for the [Gly18C6] complex in H_2O-EtOH mixed solvents at 298.15 K

X_{EtOH}, mol.fr.	0	0.12	0.25	0.50	0.74	0.91	1.0
lgK^o	0.63±0.02[8] 0.73±0.03	1.20 ±0.05	1.64 ±0.05	2.29 ±0.07	3.01 ±0.08	3.50 ±0.10	3.81 ± 0.12 [11]

Table 2. Stability constant (lgK^o) for the [Gly18C6] complex in H_2O-MeAc mixed solvents at 298.15 K

X_{MeAc}, mol.fr.	0.08	0.14	0.21
lgK^o	1.02±0.05	1.29±0.05	1.51±0.05

Table 3. Stability constant (lgK^o) for the [Gly18C6] complex in H_2O-DMSO mixed solvents at 298.15 K

X_{DMSO}, mol.fr.	0.1	0.2	0.25
lgK^o	1.28±0.10	1.80±0.10	1.92±0.10

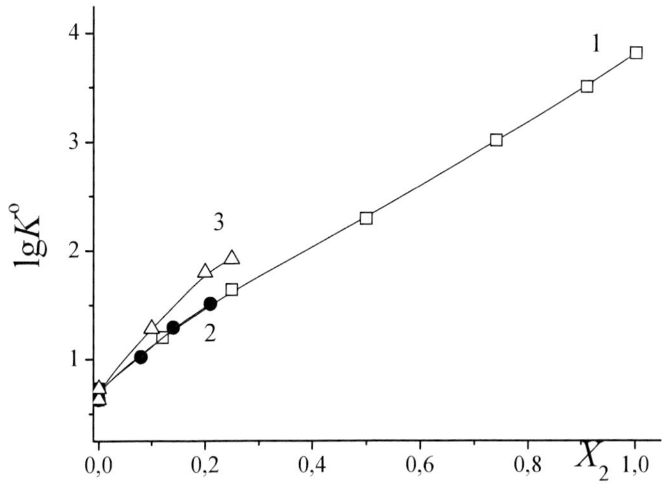

Figure 2. The comparison of the stability of the molecular complex [Gly18C6] in H_2O-EtOH [11, 16] (1), H_2O-MeAc [16, 17], (2), and H_2O-DMSO [18], (3), mixed solvents.

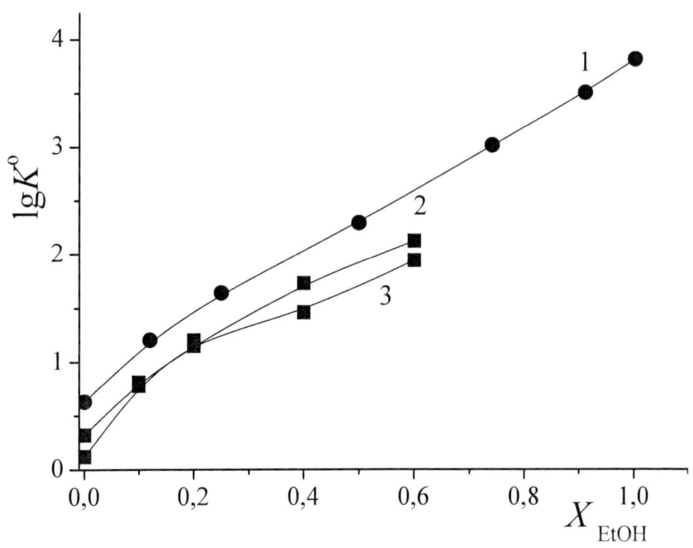

Figure 3. The comparison of the stability of the molecular complexes [Gly18C6] [11, 16], (1), [Ala18C6], [21], (2), and [Phe18C6], [22], (3), in H_2O-EtOH mixed solvents.

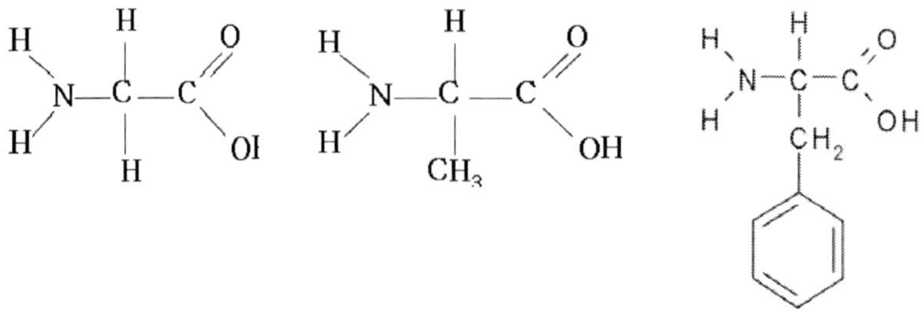

a) b) c)

Figure 4. The structure formulas of the glycine (a), alanine (b) and phenylalanine (c).

For the studied aqueous-organic solvents, the measured pH values were within physiological values (pH = 5.5 ÷ 7.0), therefore Gly was considered to be in zwitterionic form. The lg*K*°[Gly18C6] values in H_2O-EtOH [16], H_2O-MeAc [16,17] and H_2O-DMSO [18] mixed solvent are presented in tables 1-3.

The mixed solvents H_2O-EtOH, H_2O-MeAc and H_2O-DMSO influence the stability of [Gly18C6] in a similar way: the increase of the non-aqueous component concentration results in increasing lg*K*° values. The stability of [Gly18C6] is slightly higher in H_2O-DMSO solvent than in the same range of H_2O-EtOH and H_2O-MeAc solvents (figure 2).The stability of molecular complexes of glycine, *D,L*-alanine (Ala) and *L*-phenylalanine (Phe) with 18C6 in pure water slightly increases in the order lg*K*°[Phe18C6] < lg*K*°[Ala18C6] < lg*K*° [Gly18C6] (figure 3). This sequence corresponds to the increase in size of the amino acid molecules (figure 4). The benzene ring of the Phe molecule probably gives the most steric hindrance for binding thus causing the least stability of [Phe18C6]. With the addition of EtOH to water, the amino acid complex stability follows the order (lg*K*°[Phe18C6] ≈ lg*K*°[Ala18C6]) < lg*K*° [Gly18C6]. This phenomenon is presented in all range of the studied H_2O-EtOH compositions. Moreover, the result lg*K* °[Ala18C6]) < lg*K*° [Gly18C6] is also found in H_2O-MeAc and H_2O-DMSO mixtures (figure 5). The comparison of [Ala18C6] and [Gly18C6] stability shows that the difference lg*K*°[Gly18C6]) - lg*K*°[Ala18C6] is a constant for each studied solvent: 0.26±0.03, 0.34±0.05 and 0.51±0.05 logarithmic units in H_2O-EtOH, H_2O-MeAc and H_2O-DMSO mixtures, respectively. These values can be considered as the contributions of the methyl group to the [Ala18C6] stability. The analogous contribution of phenyl group to the stability of complex [Phe18C6] in H_2O-EtOH solvent is almost identical with that of the methyl group to lg$K$°[Ala18C6], although these two functional groups have different size and chemical nature.

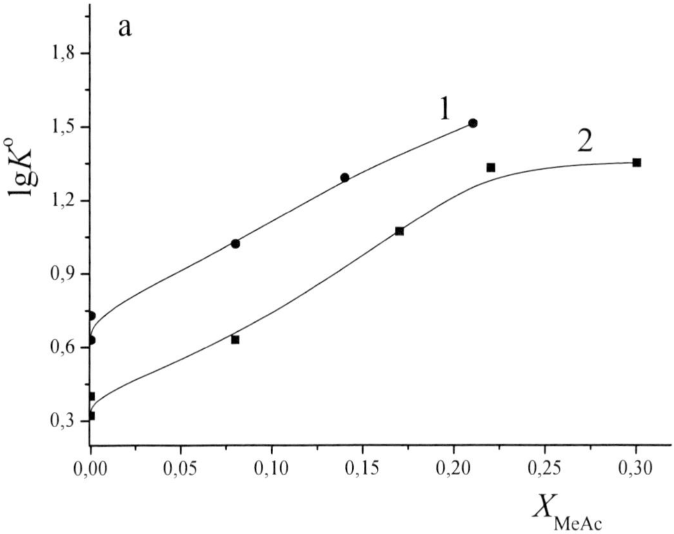

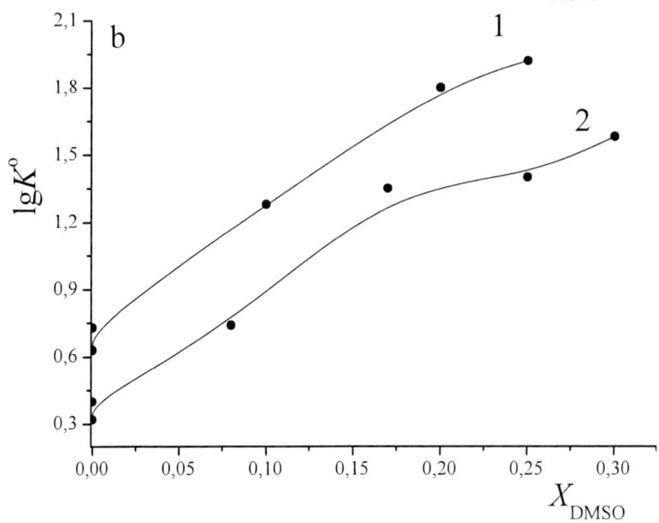

Figure 5 a,b. The stability of the molecular complexes [Gly18C6], [16 -18], (1), and [Ala18C6], [19, 20], (2), in H_2O-MeAc (a) and H_2O-DMSO (b) mixed solvents.

Figure 5a with kind permission from Springer Science+Business Media B.V.,: Ref.19, "Thermodynamic Characteristics of Alanine-18-Crown-6 Ether Complexes in binary Water-Acetone Solvents" p.36.

THERMODYNAMIC PARAMETERS OF THE [GLY18C6] MOLECULAR COMPLEX FORMATION IN H_2O-ETOH, H_2O-MEAC AND H_2O-DMSO SOLVENTS

The Gibbs energy values of the reaction (2) were calculated from eq. (7) using the data of lgK^o [Gly18C6] reported in tables 1-3.

$$\Delta_r G^\circ = -2.303\ RT\ lgK^o \quad (7)$$

These data, joined with the enthalpy of complex formation (2), $\Delta_r H^o$([Gly18C6]), obtained together with lgK^o([Gly18C6]) from calorimetric titration experiments (table 4), were used for calculating the entropy term, $T\Delta_r S^\circ$, according to eq. (8):

$$\Delta_r G^\circ = \Delta_r H^o - T\Delta_r S^\circ \quad (8)$$

Table 4. The standard thermodynamic parameter for the complex formation reaction (2) in H_2O-EtOH [16], H_2O-MeAc [16, 17] and H_2O-DMSO [18] solvents at 298.15 K: the enthalpy, $\Delta_r H^\circ$, the Gibbs energy, $\Delta_r G^\circ$, and the entropy term $T\Delta_r S^\circ$

	$-\Delta_r H^\circ$ kJ/mol	$-\Delta_r G^\circ$ kJ/mol	$-T\Delta_r S^\circ$ kJ/mol
H_2O-EtOH			
0.00	10.83±0.08 [8] 7.3±0.2	4.5 [8] 4.2	3.2 [8] 3.1
0.12	19.7±0.3	6.8	12.9
0.25	21.5±0.3	9.3	12.2
0.50	25.1±0.4	13.0	12.1
0.74	29.1±0.5	17.2	11.9
0.91	41.0±0.6	19.9	21.1
1.0[11]	64.95±0.98	21.71	43.24
H_2O-MeAc			
0.08	16.0 ±0.4	5.81	10.19
0.14	21.0 ±0.5	7.35	13.65
0.21	23.5 ±0.5	8.60	14.90
H_2O-DMSO			
0.10	17.3±1.0	7.3	10.0
0.20	26.5±1.0	10.3	16.2
0.25	30.0±1.0	11.0	19.0

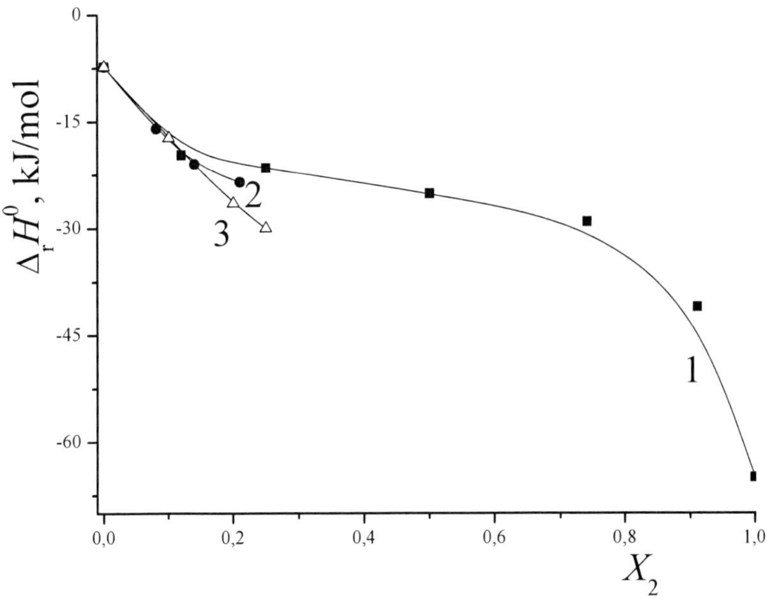

Figure 6. The influence of mixed solvents H_2O–EtOH, [11, 16], (1), H_2O–MeAc, [16, 17], (2) and H_2O-DMSO, [18], (3), on the enthalpy of the complex formation between 18C6 and Gly.

The standard thermodynamic parameter for the complex formation reaction (2) in H_2O-EtOH, H_2O-MeAc and H_2O-DMSO solvents are collected in table 4.

A similar influence of H_2O–EtOH, H_2O–MeAc and H_2O-DMSO solvents on the thermodynamic parameters of reaction (2) is found. The exothermicity of complex formation reaction increases at increasing the concentration of the non-aqueous components in all mixed solvents (figure 6). In water–ethanol mixtures this increase is more evident at X_{EtOH} = 0.91 and X_{EtOH} = 1.0 mole fractions.

The entropy change values are negative and do not favor the stability of the complex in the whole range of solvent compositions. Therefore, it can be concluded that the complex stability in H_2O–EtOH, H_2O–MeAc and H_2O-DMSO mixed solvent is determined by the enthalpy changes.

According to the solvation-thermodynamics approach [6, 46, 55-60], the influence of the solvent composition on the complex formation equilibrium can be analyzed by using the transfer thermodynamic functions, $\Delta_{tr}Y°_r$, for the Gibbs energy, $\Delta_{tr}G°_r$, the enthalpy, $\Delta_{tr}H°_r$, and the entropy $T\Delta_{tr}S°_r$.

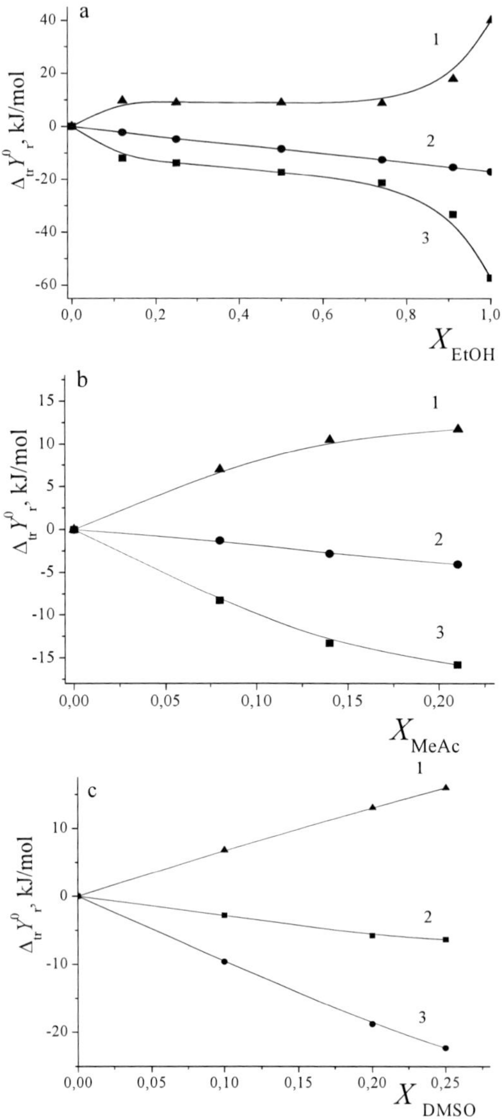

Figure 7 a-c. The thermodynamic transfer functions $-T\Delta_{tr}S°_r$, (1), $\Delta_{tr}G°_r$,(2), and $\Delta_{tr}H°_r$,(3), for the [Gly18C6] complex formation reaction in H_2O–EtOH [11, 16], H_2O–MeAc [16, 17] and H_2O-DMSO [18] mixed solvents.

Figure 7a,b with kind permission from Springer Science+Business Media B.V.: Ref.16 "Thermodynamics of complex formation in mixed solvents K and ΔH for formation reaction of [Gly18C6] at 298.15 K", p.811.

In particular, the transfer thermodynamic parameters for reaction (2) were calculated according to eq. (3) by using the thermodynamic data in mixed solvent (s) and in H_2O from table 4 as follow:

$$\Delta_{tr}G°_r = \Delta_r G°_{(s)} - \Delta_r G°_{(H2O)} \quad (9)$$

$$\Delta_{tr}H°_r = \Delta_r H°_{(s)} - \Delta_r H°_{(H2O)} \quad (10)$$

$$T\Delta_{tr}S°_r = T\Delta_r S°_{(s)} - T\Delta_r S°_{(H2O)} \quad (11)$$

The thermodynamic transfer functions are plotted as a function of solvation composition in figure 7.

The enthalpy - entropy compensation effect in the complex formation (2) was found in all solvents (figure 7 a-c). Previously, an enthalpy - entropy correlation had been found for complexation of amino acids with 18C6 and with cryptand 222 in methanol and ethanol [11]. The changes in the thermodynamic functions of the reaction (2) are the results of the changes in the solvation state of 18C6, Gly and [Gly18C6] as solvation composition changes. Thus, to explain the dependences $\Delta_{tr}Y°_r = F(X_2)$ it is necessary to analyze the influence of the mixed solvent compositions on the solvation thermodynamic parameters of the reagents and complex.

Analysis of the Influence of the Solvation Contributions of Reagents on the Gibbs Energy of Transfer of Reaction (2) from Water to H_2O-EtOH and H_2O-DMSO Solvents

The dependence of the Gibbs energy changes of reaction (2) on the Gibbs energy change for reagents solvation can be quantitatively represented by eq. (12) which follows from the thermodynamic cycle and eq. (5)

$$\Delta_{tr}G°_r = \Delta_{tr}G°([Gly18C6]) - \Delta_{tr}G°(Gly) - \Delta_{tr}G°(18C6) \quad (12)$$

where $\Delta_{tr}G°([Gly18C6])$, $\Delta_{tr}G°(Gly)$ and $\Delta_{tr}G°(18C6)$ are the changes of Gibbs energy of solvation for complex [Gly18C6], Gly and 18C6 as the solvent composition changes.

The quantities in the right-hand side of eq.(12) were calculated as follows.

Table 5. Gibbs energy transfer of Gly, $\Delta_{tr}G°$(Gly), from H_2O into H_2O-EtOH solvents at $T = 298.15$ K

$X_{(EtOH)}$ mol. fr.	0.03	0.07	0.12	0.16	0.24	0.31	0.42	0.55	0.74	1.00
$\Delta_{tr}G°$(Gly) [21, 22]	0.17	0.36	2.38	2.63	3.20	7.64	8.68	8.74	11.15	14.79 19.7*

*calculated from solubility data of Gly in EtOH reported in [11].

The $\Delta_{tr}G°$(Gly) values for H2O-EtOH mixtures were obtained from solubility data of Gly in water-ethanol mixtures according to:

$$\Delta_{tr}G°(\text{Gly}) = -2.303\ RT \log \text{Cs/Cw} \qquad (13)$$

where Cs, Cw are the solubility of Gly in H_2O-EtOH solvents and in water, respectively. In the literature a set of solubility data of amino acids in alcohols [11] and their binary mixtures with water [61, 62] can be found. The data on $\Delta_{tr}G°$ of amino acids reported in the table 5 were obtained by eq. 13 using the Gly solubility data in the whole range of water-EtOH solvent composition (0.0-1.0 mole fraction of EtOH) reported in [61]. The calculation results agree with the $\Delta_{tr}G°$(Gly) values reported in the range X (H_2O-EtOH) = 0.0 ÷ 0.3 mole fraction of EtOH in [62] and with $\Delta_{tr}G°(\text{Gly})_{(H2O\text{-}EtOH)}$ value for EtOH [11]. Values of $\Delta_{tr}G°(\text{Gly})_{(H2O\text{-}DMSO)}$ were reported in [64]. The thermodynamic parameters $\Delta_{tr}G°(18C6)_{(H2O\text{-}EtOH)}$ and $\Delta_{tr}G°(18C6)_{(H2O\text{-}DMSO)}$ were determined by us earlier in [65] and in [66] respectively. Finally, rearrangement of eq.(12) allows the calculation of $\Delta_{tr}G°$([Gly18C6]):

$$\Delta_{tr}G°([\text{Gly18C6}]) = \Delta_{tr}G°_r + \Delta_{tr}G°(\text{Gly}) + \Delta_{tr}G°(18C6) \qquad (14)$$

The behavior of all $\Delta_{tr}G°$'s, shown in figure 8a,b, suggests the following observations:

1. The desolvation of Gly plays a prevailing role in all studied range of H_2O-EtOH and H_2O-DMSO solvents.
2. The values $\Delta_{tr}G°_r$ correspond to the values $\Delta_{tr}G°$(Gly) but are of opposite sign.
3. The changes in the solvation state of [Gly18C6], Δ_{tr}G°([Gly18C6]), are small and close to the change in the solvation state of 18C6,

$\Delta_{tr}G°(18C6)$. This fact points out the determining role of the macrocycle in the formation of the solvation shell of the complex: it can be assumed that the change in the solvation state of carboxylic group of Gly, which does not participate in coordination with oxygen atoms of crown-ring, has no influence on the change in solvation state of [Gly18C6].

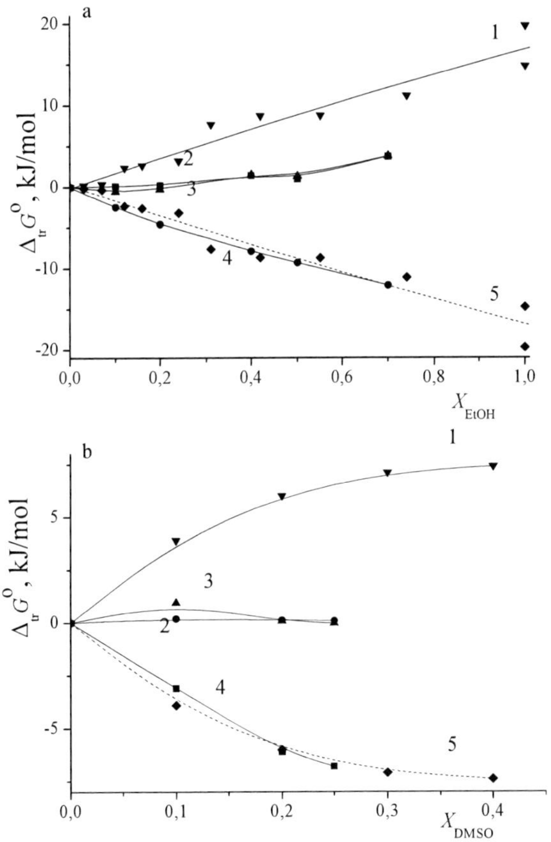

Figure 8 (a, b). The solvation contributions of reagents and complex to the changes of Gibbs energy of the reaction of the [Gly18C6] molecular complex formation in H_2O-EtOH (a) and in H_2O -DMSO (b) solvents. $\Delta_{tr}G°(Gly)$ (1), $\Delta_{tr}G°(18C6)$ (2), $\Delta_{tr}G°([Gly18C6])$ (3), $\Delta_{tr}G°_r$(4), $-\Delta_{tr}G°(Gly)$ (5).

Figure 8 a,b with kind permission from Springer Science+Business Media B.V.: Ref.21 "Calorimetric investigation of the complex formation reaction of 18-crown-6 ether with d,l-alanine in water–ethanol mixtures".

Analysis of the Influence of the Solvation Contributions of Reagents on the Enthalpy of Transfer of Reaction (2) from Water to H_2O-EtOH, H_2O-MeAc and H_2O-DMSO Solvents

Eq. (15), which is derived from (5), was applied for the solvation-thermodynamic analysis of the enthalpy solvation contributions of Gly, ($\Delta_{tr}H°$(Gly)) and 18C6, ($\Delta_{tr}H°$(18C6)) to the reaction enthalpy transfer from water to H_2O-EtOH, H_2O-MeAc and H_2O-DMSO solvents:

$$\Delta_{tr}H°_r=\Delta_{tr}H°([Gly18C6]) - \Delta_{tr}H°(Gly) - \Delta_{tr}H°(18C6) \quad (15)$$

The $\Delta_{tr}H°_r$ values were determined from equation (10) using the data of table 4. The enthalpy characteristics of the solvation of reagents in the studied mixed solvents were reported in literature, in particular for Gly in H_2O-EtOH [67], in H_2O-MeAc [68], and in H_2O-DMSO [69] mixtures. The enthalpic characteristics of 18C6 were presented in [42] for H_2O-EtOH, in [17] for H_2O-MeAc and in [41] for H_2O-DMSO solvents.

The enthalpy of the [Gly18C6] complex transfer, ($\Delta_{tr}H°$([18C6])), from water to mixed solvents was calculated according to:

$$\Delta_{tr}H°([Gly18C6]) = \Delta_{tr}H°_r + \Delta_{tr}H°(Gly) + \Delta_{tr}H°(18C6) \quad (16)$$

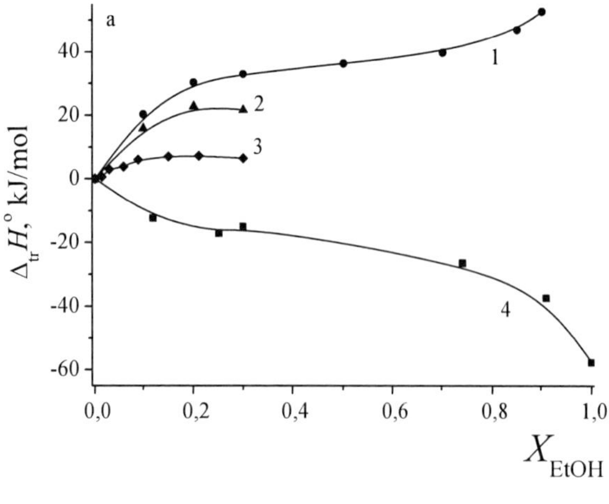

Figure 9. (Continued).

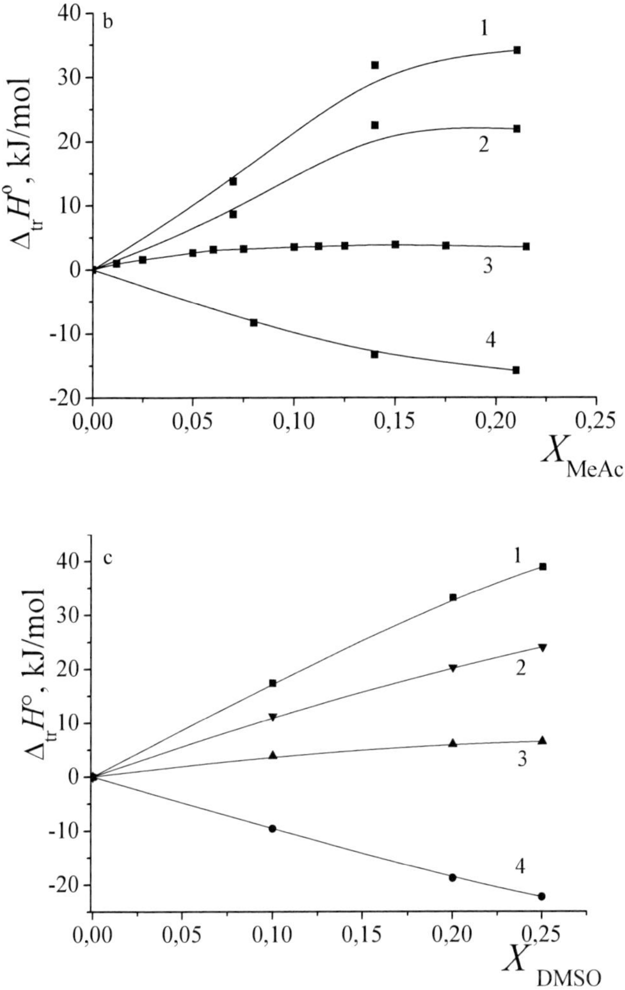

Figure 9 a-c. The influence of H_2O-EtOH (a), H_2O-MeAc (b) and H_2O-DMSO (c) mixed solvent on the enthalpic characteristics of reaction of [Gly18C6] molecular complex formation. $\Delta_{tr}H°$(18C6) [42, 17, 41]; 2-$\Delta_{tr}H°$([Gly18C6]); 3-$\Delta_{tr}H°$(Gly) [67-69]; 4-$\Delta_{tr}H°_r$ [16-18].

Figure 9a with kind permission from Springer Science+Business Media B.V.: Ref.16 "Thermodynamics of complex formation in mixed solvents K and ΔH for formation reaction of [Gly18C6] at 298.15 K" p.811.

All these quantities are plotted in figure 9a,b. This figure clearly indicates that the desolvation of 18C6 and Gly with increasing of the concentration of EtOH, MeAc and DMSO in the mixed solvents favors the increase of the exothermicity of the reaction. The solvation contribution of 18C6, $\Delta_{tr}H°$(18C6), prevails over the small and insignificant Gly contribution, $\Delta_{tr}H°$(Gly), and over the [Gly18C6] molecular complex contribution, $\Delta_{tr}H°$([Gly18C6]).

Comparison of Solvation Contribution of Reagents to the Gibbs Energy Change and to the Enthalpy Change of Transfer from Water to H_2O-EtOH and H_2O- DMSO Mixtures

The behavior of the solvation contributions of the reagents for change of Gibbs energy of reaction (2) in H_2O-EtOH and H_2O-DMSO solvents (figure 8a,b) is essentially different from the behavior of the corresponding quantities for the enthalpy change (figure 9a-c). The changes in Gly solvation state expressed in the Gibbs energy term, $\Delta_{tr}G°$(Gly), result the main factor in determining the reaction Gibbs energy as the solvation composition changes: $\Delta_{tr}G°_r \approx - \Delta_{tr}G°$(Gly). On the contrary, the analysis of enthalpy contributions of reagents, $\Delta_{tr}H°$(Gly) and $\Delta_{tr}H°$(18C6), has shown a dependence of $\Delta_{tr}H°_r$ on $\Delta_{tr}H°$(18C6), that is, $\Delta_{tr}H°_r = F(\Delta_{tr}H°(18C6))$.

At the same time, the $\Delta_{tr}H°$(Gly) and $\Delta_{tr}G°$(18C6) values are very small and negligible in comparison with the corresponding thermodynamic parameters of the reaction, $\Delta_{tr}Y°_r$, and of the complex, $\Delta_{tr}Y°$([Gly18C6]).

This phenomenon can be explained by a comparative analysis of thermodynamic solvation functions of 18C6 and Gly in H_2O-EtOH and H_2O-DMSO solvents (figure 10 a,b).

In both solvents, the Gibbs energy term of Gly, $\Delta_{tr}G°$(Gly), changes more rapidly than $\Delta_{tr}G°$(18C6) as the non aqueous component in solvents increases. Instead, the influence of mixed solvent composition is more evident in the enthalpy solvation characteristics of 18C6, $\Delta_{tr}H°$(18C6), in comparison with $\Delta_{tr}H°$(Gly).

Probably the determining role of the Gibbs energy contribution of glycine, $\Delta_{tr}G°$(Gly), in the Gibbs energy change of the reaction, $\Delta_{tr}G°_r$, is due to the more dynamic changes of $\Delta_{tr}G°(Gly)$ in comparison with the Gibbs energy transfer of crown ether $\Delta_{tr}G°(18C6)$. The opposite occurs for the enthalpy contributions of Gly, $\Delta_{tr}H°(Gly)$, and 18C6, $\Delta_{tr}H°(18C6)$.

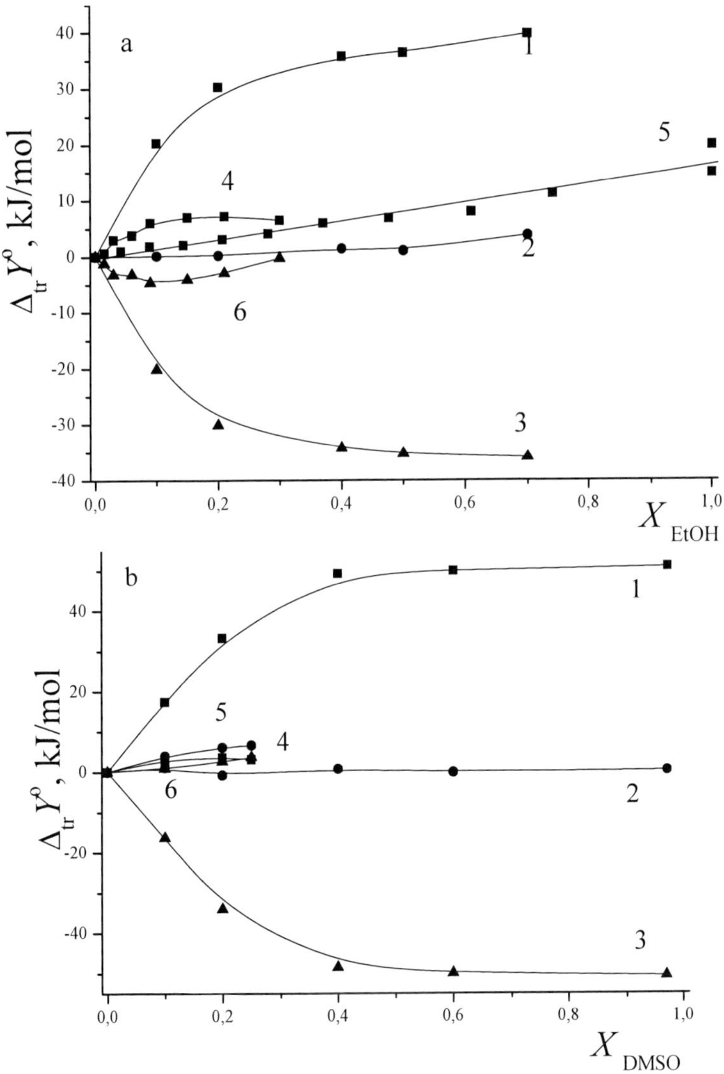

Figure 10 a,b. Solvation transfer function for Gly and 18C6 in H_2O-EtOH (a) and in H_2O-DMSO (b) solvents. 1-$\Delta_{tr}H°(18C6)$ [41, 42]; 2-$\Delta_{tr}G°(18C6)$ [65, 66]; 3-$-T\Delta_{tr}S°(18C6)$; 4-$\Delta_{tr}H°(Gly)$ [67, 69]; 5-$\Delta_{tr}G°(Gly)$ [21, 22]; 5- $-T\Delta_{tr}S°(Gly)$.

Solvation-Thermodynamics Model for Predicting the Changes in the Thermodynamic Quantities of Reaction for the Formation of Ionic and Molecular Complexes of 18C6 in Mixed Solvents

The key role of the ligand solvation contribution to the thermodynamics of complexation was established earlier for reactions of the formation of amine and carboxylate complex of ions of *d*-metals in water–organic solvents [46, 55–60], but for these reactions the solvation state of the complex ion was mainly determined by the solvation state of the complexation ion. In the reaction of formation of the molecular complex, glycine formally plays the role of the complexation ion. During the formation of crown complexes, the solvation shell of the complex particle is mainly formed by the solvation shell of the macrocycle, which to a considerable extent screens the complexation ion from the effect of the solvent. This is indicated by the very similar trends of the enthalpies of solvation of 18C6 and of the molecular complex as the DMSO concentration in the mixed solvent increases.

A study of complex formation reaction between Ni^{+2} and amines in water-organic mixed solvents [45] showed not only that the changes in the thermodynamic quantities of reaction, $\Delta_{tr}Y_r°$, depend on the change of the parameters of the ligand solvation, but also that the $\Delta_{tr}Y_r°$ values are in the range of changes of the corresponding transfer function of the ligand, $\Delta_{tr}Y°(L)$, and have opposite sign:

$$0 < |\Delta_{tr}Y_r°| < |\Delta_{tr}Y°(L)| \qquad (17)$$

Further studies [46, 55, 56] found that this dependence is more general and is manifested not only in the formation of amine complexes of Ni^{+2}, but also in the complex formation reactions of Cu^{+2}, Zn^{+2}, Cd^{+2} and Ag^{+} with the amines and carboxylate-type ligands. This pattern has been named as "rule of ranges" [46, 55, 56]. Moreover, the solvation effects of ions, expressed as the difference between the change in solvation characteristics of the central ion and its complex, $\Delta_{tr}Y°([ML]^{+})$ - $\Delta_{tr}Y°(M^{+})$, resulted proportional to the changes in the ligand solvation parameters, $\Delta_{tr}Y°(L)$, and did not exceed its values. The coefficient of proportionality α_{dif}, also named as "coefficient of differences" can be calculated as

$$\alpha_{dif} = (\Delta_{tr}Y^{o}([ML]^{+}) - \Delta_{tr}Y^{o}(M^{+}))/\Delta_{tr}Y^{o}(L) \quad (18)$$

This parameter depends on the nature of the metal and the ligand and its values fall in the range $0 < \alpha_{dif} < 1$, and remain approximately constant over the entire range of compositions of binary solvent. It describes a quantitative relationship between the functions of solvation of the individual reagents: the greater the relative differences in the solvation states of the metal ion and the complex, the greater the coefficient of differences.

The values of α_{dif} are tabulated for some *d*-metal ions and certain types of ligands. An analysis of a significant volume of experimental data on thermodynamic characteristics of Ni^{+2}, Cu^{+2}, Zn^{+2}, Cd^{+2} and Ag^{+} ion complexes with amines and carboxylate ligands in aqueous-organic mixed solvents has shown that for these reactions the α_{dif} values fall in the range 0.6 - 0.8 [46, 55, 56].

The great opportunity offered by α_{dif} is that the knowledge of the changes in thermodynamic properties of ligand allows to predict the changes in stability of the complexes and the heat effects of complex formation reactions in different media by using a set of equations:

$$\Delta_{tr}Y^{o}_{r} = \{\Delta_{tr}Y^{o}([ML]^{+}) - \Delta_{tr}Y^{o}(M^{+})\} - \Delta_{tr}Y^{o}(L) = \alpha_{dif}\Delta_{tr}Y^{o}(L) - \Delta_{tr}Y^{o}(L) \quad (19)$$

or

$$\Delta_{tr}Y^{o}_{r} = (\alpha_{dif} - 1)\,\Delta_{tr}Y^{o}(L) \quad (20)$$

where $\Delta_{tr}Y^{o}$ are the transfer thermodynamic functions ($\Delta_{tr}H^{o}$, $\Delta_{tr}G^{o}$, $T\Delta_{tr}S^{o}$) for the reaction ore reagents solvation.

The thermodynamic relationships described above for *d*-metal complex formation with amines and carboxylate ligands also occur In the cases of complex formation between 18C6 and Ag^{+} in H_2O – DMSO and H_2O – EtOH mixed solvents [41,42]. In particular, the differences in the enthalpy change of solvation of the central ion, $\Delta_{tr}H^{o}(Ag^{+})$, and the complex, $\Delta_{tr}H^{o}([Ag18C6]^{+})$, does not exceed the enthalpy change of transfer of the ligand, $\Delta_{tr}H^{o}(18C6)$, therefore the coefficient of differences can be calculated as:

$$\alpha_{dif} = (\Delta_{tr}H^{o}([Ag18C6]^{+}) - \Delta_{tr}H^{o}(Ag^{+}))/\Delta_{tr}H^{o}(18C6) \quad (21)$$

Table 6. The influence of H_2O – DMSO solvents on the enthalpic terms of complex formation reaction between 18C6 and silver (I) ions at T=298.15K

X_{DMSO} mol.fr	$\Delta_{tr}H°_r$ kJ/mol [41]	$\Delta_{tr}H°([Ag18C6]^+)$- $\Delta_{tr}H°(Ag^+)$ * ±4.5kJ/mol	$\Delta_{tr}H°(18C6)$ ±0.3kJ/mol [41]	α_{dif}
0.1	-5.3±1.7	12.1	17.4	0.7
0.2	-11.0 ± 2.5	22.2	33.2	0.7
0.3	-13.6 ± 2.8	31.0	44.6	0.7
0.4	-12.8 ± 2.8	36.3	49.2	0.7
0.6	-8.0 ± 2.8	41.9	49.9	0.8
0.7	-9.6 ± 3.2	40.5	50.0	0.9
0.8	-7.0 ± 3.5	43.5	50.5	0.9
0.97	-2.8 ± 3.5	48.1	50.9	0.9

*$\Delta_{tr}H°(Ag^+)$ are from [70].

Table 7. The influence of H_2O – EtOH solvents on the enthalpic terms of complex formation reaction between 18C6 and silver (I) ions at T=298.15K

X_{EtOH} mol.fr	$\Delta_{tr}H°_r$ ± 0.2 kJ/mol [42]	$\Delta_{tr}H°([Ag18C6]^+)$- $\Delta_{tr}H°(Ag^+)$ * ± 4.5 kJ/mol	$\Delta_{tr}H°(18C6)$ ± 0.2 kJ/mol [42]	α_{dif}
0.1	-0.8	19.6	20.4	0.9
0.2	-2.0	28.3	30.3	0.9
0.3	-2.2	30.7	32.9	0.9
0.5	-3.8	32.5	36.3	0.9
0.7	-7.4	32.3	39.7	0.8
0.9	-17.3	35.4	52.7	0.7

* $\Delta trH°(Ag^+)$ are from [70]

The average α_{dif} value in both H_2O – DMSO and H_2O – EtOH solvent is 0.8 ± 0.1 (tables 6, 7). This allows using eq (22) to estimate the heat effects of reaction of complex formation between 18-crown-6 and silver (I) ion in aqueous-organic solvents:

$$\Delta_{tr}H°_{r,\,calc} = (\alpha_{dif} - 1)\,\Delta_{tr}H°(18C6) \qquad (22)$$

Figure 11 a,b illustrates that eq. (22) satisfactorily describes the experimental results.

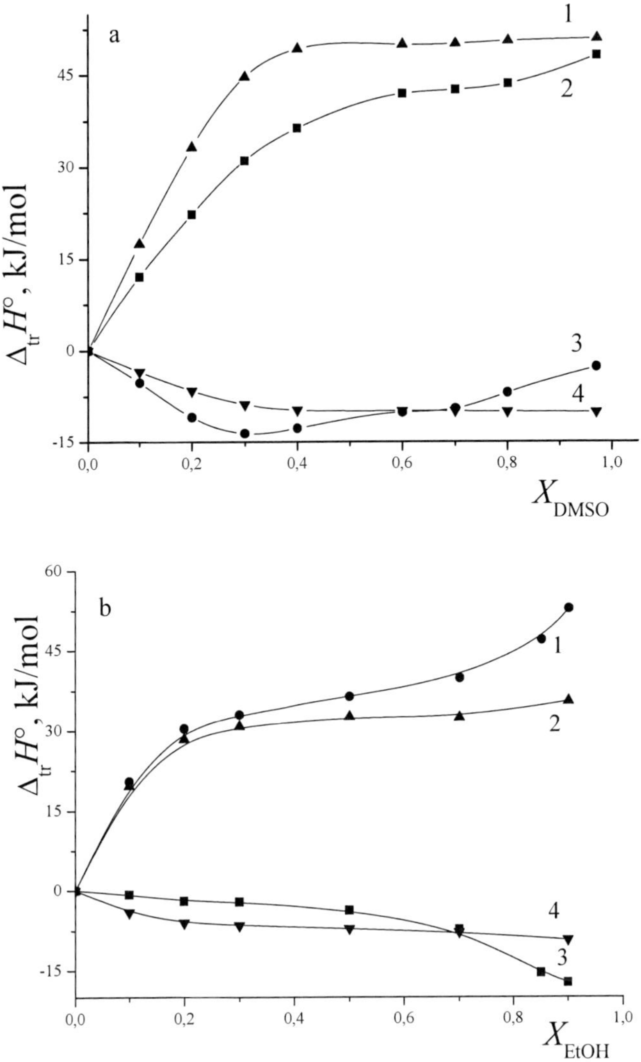

Figure 11 a,b. Experimental, $\Delta_{tr}H^{o}{}_{r}$, and calculated by eq. 22, $\Delta_{tr}H^{o}{}_{r,\,calc}$, enthalpic characteristics of the $[Ag18C6]^{+}$ complex formation reaction in H_2O – DMSO (a) and H_2O – EtOH (b) solvents, $\alpha_{dif} = 0.8$. 1- $\Delta_{tr}H^{o}(18C6)$ [41, 42, 70]; 2- $\{\Delta_{tr}H^{o}([Ag18C6]^{+})$- $\Delta_{tr}H^{o}(Ag^{+})\}$; 3- $\Delta_{tr}H^{o}{}_{r}$ [41, 42,70]; 4- $\Delta_{tr}H^{o}{}_{r,\,calc}$

The applicability of the "rule of ranges" was analyzed also for the enthalpic characteristics of [Gly18C6] molecular complex formation reaction in H_2O–EtOH, H_2O–MeAc, and H_2O–DMSO solvents.

The difference between the enthalpy changes in the solvation state of the molecular complex, and Gly, $\Delta_{tr}H°([Gly18C6]) - \Delta_{tr}H°(Gly)$, is proportional to the changes in the solvation state of the 18C6, $\Delta_{tr}H°(18C6)$ (figure 9 a-c), with α_{dif} as proportionality coefficient [46, 55, 56]:

$$\Delta_{tr}H°([Gly18C6]) - \Delta_{tr}H°(Gly) = \alpha_{dif}\,\Delta_{tr}H°(18C6) \qquad (23)$$

The α_{dif} values calculated from eq. (23) in the H_2O–EtOH, H_2O–MeAc, and H_2O–DMSO solvents are 0.45±0.01 and do not depend on the solvents nature and compositions. The value of α_{dif} = 0.45 has been used in equation (22) for an estimation of the enthalpy changes of the formation of [Gly18C6] complex.

The experimental values, $\Delta_{tr}H°_r$, and calculated according to eq. (22), $\Delta_{tr}H°_{r\cdot calc}$, compare well with each other (figure 12 a-c). The calculated values, $\Delta_{tr}H°_{r.calc}$, allow estimating the influence of H_2O-EtOH, H_2O-MeAc and H_2O-DMSO solvent composition on the energetics of complex formation between 18C6 and Gly at low solubility of glycine in the solvents with high concentration of non aqueous components, where the experimental determination of the reaction enthalpy is impossible now.

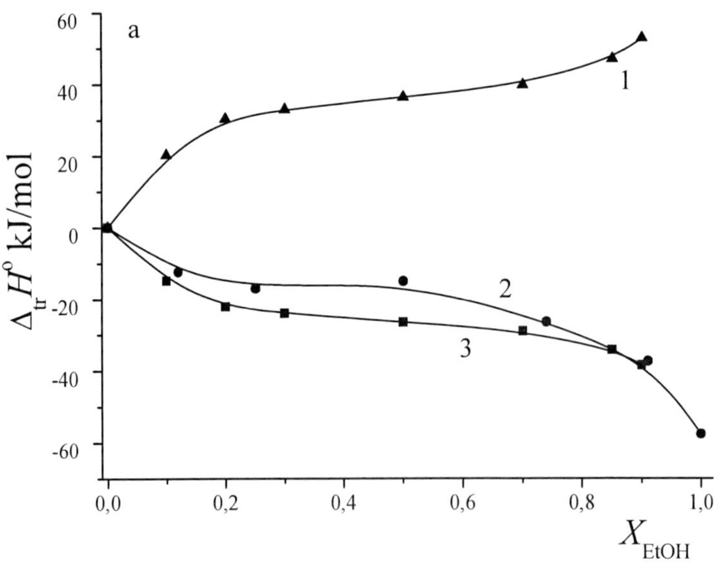

Figure 12. (Continued).

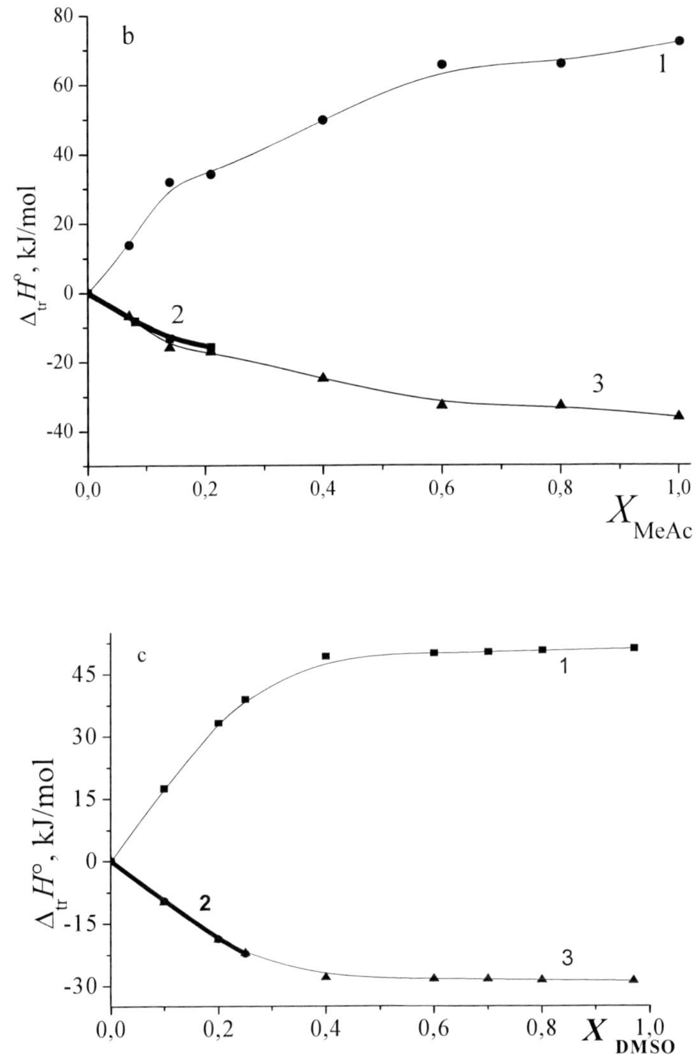

Figure 12 a-c. Enthalpic solvation characteristics for 18C6, $\Delta_{tr}H^o(18C6)$ (1) in mixed solvents H_2O-EtOH [42], (a), H_2O-MeAc [17], (b), H_2O- DMSO [41], (c); experimental, $\Delta_{tr}H^o_r$, [18] (2), and calculated, $\Delta_{tr}H^o_{r,calc.}$, [18], (3), enthalpic characteristics for the reaction of the [Gly18C6] molecular complex formation.

CONCLUSION

In comparison with the ionic complexes [46, 55, 56], the α_{dif} values for [Gly18C6] molecular complex formation reaction are lower. They were obtained by grouping the enthalpy contribution terms of reagents in the same way as for the enthalpic contributions of ionic complexes. Such combination was based on the similarity of the thermodynamic characteristics of solvation: the solvation shell of complex ions is determined by the changes in solvation of the complex particle as the solvent composition changes.

As it was inferred from the studies on *d*-metal complex formation [46, 55, 56], the α_{dif} was a sort of measure of the degree of the difference between the thermodynamic properties of solvent molecules and ligand in their solvation action towards the ion, because the reaction of complex formation is at the same time the reaction of substitution of solvent molecules by the molecule(s) of ligand in the first solvation shell of the ion. A value of α_{dif} near 1 means a significant thermodynamic difference between the solvating properties of the ligands and solvent molecules. Inversely, a small value of the α_{dif} characterizes a thermodynamic similarity between molecules of solvent and ligand.

In consequence, the α_{dif} value obtained for [Gly18C6] complex formation (0.45±0.01) appears smaller than expected, that is, as in the case of $[Ag18C6]^+$ complex formation (0.8).

The reason of this anomaly is not clear now. We can suggest that in the cases of [Gly18C6] complex formation the lower value of α_{dif} is related to the complex formation features of 18C6 as macrocyclic ligand and Gly in addition to the differences in the thermodynamic properties of solvent molecules and molecules of one of the reagents.

The application of the features established previously for *d*-metal complex formation [46, 55, 56] to the thermodynamics of the formation of the molecular complex of Gly with 18C6 in H_2O-EtOH, H_2O-MeAc and H_2O-DMSO mixed solvents is illustrated in figure 12 a-c.

These features can be used for estimating the change in the enthalpy of the formation of the molecular complex in solvents with high acetone or DMSO concentrations, in which the experimental determination of the enthalpy characteristics of complexation is hampered by the low solubility of glycine.

REFERENCES

[1] Dietrich, B.; Viout, P.; Lehn, J.M. Macrocyclic chemistry; VCH Verlagsgesellschaft mbH., Weinheim, GERMANY (GE), 1993; pp 1-321.

[2] Lehn, J.M. Supramolecular chemistry. Concepts and perspectives; VCH Verlagsgesellschaft mbH., Weinheim, GERMANY (GE), 1995; pp 1-334.

[3] Schneider, H.J.; Yatsimirsky, A.K. Principles and methods in supramolecular chemistry, Wiley, Chichester, U.K., 2000; pp 137–190.

[4] Steed, J.W.; Atwood, J.L. Supramolecular Chemistry; Academkniga, Moscow, RUSSIA (RU), 2007; Vol. 1, pp 1-480.

[5] Terekhova, I.V. In Macrocyclic Chemistry: New Research Developments; Fitzpatrick, D.W; Ulrich, H.J.; Ed.; Nova Science Publisher: New York, USA, 2010; pp 263-286.

[6] Krestov, G.A. Ionic Solvation; Ellis Horwood, New-York-London-Toronto-Sydney-Tokyo-Singapore, 1994; pp 1-264.

[7] Terekhova, I.V.; Parfenyuk, E.V.; Kulikov, O.V. *J. Therm. Anal. Calorim.,* 2002, vol 68, 185-189.

[8] Kulikov, O.V.; Lapshev, P.V.; Terekhova, I.V. *Rus. J. Phys. Chem.,* A. 1998, vol 72, 632-636.

[9] Kulikov, O.V; Terekhova, I.V; *Rus. J. Coord Chem.,* 1998, vol 24, 373-378.

[10] Danil de Namor, A.F.; Ritt, M.C.; Lewis, D.F.V.; Schwing –Weill, M.J.; Neu, F.A. *Pure Appl. Chem.,* 1991, vol 63, 1435-1439.

[11] Danil de Namor, A.F.; Ritt, M.C.; Schwing-Weill, M.J.; Arnaud-Neu, F.; Lewis, D.F.V. *J. Chem. Soc. Faraday Trans.,* 1991, *vol* 87, 3231-3239.

[12] Buschmann, H-J.; Mutihac, L. *J. Inclus. Phenom. Macrocyclic Chem.* 2002, *vol* 42, 193–195.

[13] Mutihac, L.; Buschmann, H-J.; Jansen, K.; Wego, A. *Materials Science Engineering,* C. 2001, vol 18, 259–264.

[14] Czekalla, M.; Stephan, H.; Habermann, B.; Trepte, J.; Gloe, K.; Schmidtchen, F.P. *Thermochim. Acta.* 1998, *vol* 313, 137-144.

[15] Buschmann, H.J.; Schollmeyer, E.; Mutihac, L. *Thermochim. Acta.,* 1998, vol 316, 189-192.

[16] Matteoli, E.; Lepori, L.; Usacheva, T.R.; Sharnin, V.A. *J. Therm. Anal. Calorim.,* 2009, vol 97, 811-816.

[17] Usacheva, T.R.; Kuz'mina, I.A.; Sharnin, V.A.; Chernov, I.V.; Matteoli, E. *Rus. J. Phys. Chem.,* A. 2011, vol 85, 948–951.

[18] Usacheva, T.R.; Sharnin, V.A.; Matteoli, E. *Rus. J. Phys. Chem.,* A. 2011, vol 85, 1898–1902.
[19] Usacheva, T.R.; Kuz'mina, I.A.; Sharnin, V.A.; Chernov, I.V.; Matteoli, E. *Rus. J. Phys. Chem.,* A. 2012, vol 86, 36–39.
[20] Usacheva, T.R.; Kuz'mina, I.A.; Sharnin, V.A.; Chernov, I.V.; Matteoli, E. *Rus. J. Phys. Chem.,* A. 2012, vol 86, 1064-1067.
[21] Usacheva, T.R.; Sharnin, V.A.; Chernov, I.V.; Matteoli, E. *J. Therm. Anal. Calorim.,* DOI 10.1007/s10973-012-2625-7.
[22] Usacheva, T.R.; Sharnin, V.A.; Chernov, I.V.; Matteoli, E., Terekhova, I.V.; Kumeev, R.S. *Chem. Phys. Lett.,* 2012, vol 543, 155-158.
[23] Pedersen, C.J. *J. Amer. Chem. Soc.,* 1967, vol 89, 7017-7036.
[24] Pedersen, C. J. *Angew. Chem.,* 1988, vol 100, 1053-1059.
[25] Hiraoka, M. Crown compounds. Their characteristics and applications. Mir: Moscow, RUSSIA (RU), 1986; pp 1-359.
[26] Frensdorff, H.K. *J. Amer. Chem. Soc.,* 1971, vol 93, 600-606.
[27] Izatt, R.M.; Eatough, D.J.; Christensen, J.J. *Struct. bond.,* 1973, vol 16, 161-189.
[28] Izatt, R.M.; Terry, R.E.; Haymore, B.L.; Hansen, L.D.; Dalley, N.K.; Avonted, A.G.; Christensen, J.J. *J. Amer. Chem. Soc.,* 1976, vol 98, 7620-7626.
[29] Bradshaw, J.S.; Maas, G.E.; Lamb, J.D. *J. Amer. Chem. Soc.,* 1980, vol 102, 467-475.
[30] Lamb, J.D.; Izatt, R.M.; Swain, C.S.; Christensen, J.J. *J. Amer. Chem. Soc.,* 1980, vol 102, 475-480.
[31] Buschmann, H.-J. *Chem. Ber.,* 1985, vol 118, 2746-2756.
[32] Larina, O.V.; Kern, A.P.; Bondarev, N.V. *Rus. J. Gen. Chem*., 1997, vol 67, 1439-1442.
[33] Izatt, R.M.; Lamb, J.D.; Asay, E.; Maas, E.G.; Bradshaw, J.S.; Christensen, J.J. *J. Amer. Chem. Soc.,* 1977, vol 99, 6134-6136.
[34] Lamb, J.D.; Izatt, R.M.; Swain, C.S.; Bradshaw, J.S.; Christensen, J.J. *J. Amer. Chem. Soc.,* 1980, vol 102, 480-482.
[35] Almasio, M.-C.; Arnaud-Neu, F.; Schwing-Weill, M.J. *Helv. Chim. Acta.,* 1983, vol 66, 1296-1305.
[36] Buschmann, H.J.; Schollmeyer, E. *J. Electroanal. Chem.,* 1999, vol 474, 188-191.
[37] Buschmann, H.-J.; *Polyhedron.* 1992, vol 11, 559-561.
[38] Buschmann, H.-J.; *Polyhedron.* 1987, vol 6, 1469-1472.
[39] Buschmann, H.-J.; Schollmeyer, E. *Inorg. Chem Acta.,* 2000, vol 298, 120-122.

[40] Lada, E.; Lei, X.; Kalinowski, K. *Monatsh. Chem.*, 1992, vol 123, 425-433.
[41] Usacheva, T.R.; Ledenkov, S.F.; Sharnin, V.A. *J. Therm. Anal. Calorim.*, 2002, vol 70, 209-216.
[42] Usacheva, T.R.; Ledenkov, S.F.; Sharnin, V.A. *J. Therm. Anal. Calorim.*, 2002, vol 70, 379-385.
[43] Jozwiak, M.; Blad, A.; Piekarski, H. *J. Mol. Liquids.*, 2003, vol 107, 155-167.
[44] Jozwiak, M.; Piekarski, H. *J. Mol. Liquids*, 2003, vol 106, 15-29.
[45] Shormanov, V.A. In Complex formation in non aqueous media. Krestov G.A.; Ed.; Series Problems of solution chemistry; Nauka: Moscow, RUSSIA (RU), 1989; pp 143-189.
[46] Shormanov, V.A; Sharnin, V.A. In Achievements and Problems of Solvation Theory: Structure-Thermodynamical Aspects, Kutepov, A.M.; Ed.; Nauka: Moscow, RUSSIA (RU), 1998; pp 172-205.
[47] Matteoli, E.; Mansoori, G.A. Fluctuation Theory of Mixtures, Taylor and Francis: New York, USA, 1990.
[48] Matteoli, E.; Lepori, L. *J. Chem. Soc. Faraday Trans.*, 1995, vol 91, 431-436.
[49] Marcus, Y. *Monatsh. Chem.*, 2001, vol, 132, 1387-1411.
[50] Marcus, Y. *Pure Appl. Chem.*, 1990, vol 62, 2069-2076.
[51] Marcus, Y. *J. Solut. Chem.*, 2007, vol 36, 1385-1399.
[52] Contrani, L.; Mennucci, B.; Tomasi, J.; *J. Mol. Struct. Theochem.*, 2000, vol 500, 113-127.
[53] Kalidas, C.; Hefter, G.; Marcus, Y. *Chem. Rev.*, 2000, vol 100, 819-852.
[54] Hefter, G.; Marcus, Y.; Waghorne, W.E. *Chem. Rev.*, 2002, vol 102, 2773-2836.
[55] Sharnin, V.A. *Russ. J. Gen. Chem.*, 1999, vol 69, 1421-1429.
[56] Sharnin, V.A.; *Izv. Vyssh. Uchebn. Zaved. Khim. Khim. Tekhnol.* 2005, vol 48, 44–53.
[57] Gesse, Zh.F.; Repkin, G.I; Isaeva, V.A; Sharnin, V.A. *J. Therm. Anal. Calorim.*, DOI 10.1007/s10973-011-2127-z
[58] Zevakin, M.A; Grazhdan, K.V.; Dushina, S.V.; Sharnin, V.A. *J. Mol. Liquid.*, 2007, vol 131-132, 163-167.
[59] Sharnin, V.A.; Dushina, S.V.; Zevakin, M.A.; Gushchina, A.S.; Grazhdan, K.V. *Inorg. Chim. Acta.*, 2009, vol. 362, 437-442.
[60] Ledenkov, S.F.; Shormanov, V.A.; Sharnin, V.A. *Rus. J. Phys. Chem.*, 1996, vol 70, 1642–1646.
[61] Dey, B.P.; Lahiri, S.C. *Indian. J. Chem.*, 1986, vol 25A, 136-140.

[62] Spink, C.H.; Auker, M. *J. Phys. Chem.,* 1970, vol 74, 1742-1747.
[63] Chakravorty, S.K.; Lahiri, S.C. *J. Indian. Chem. Soc.,* 1987, vol 64, 399-402.
[64] Gesse, Zh.F.; Isaeva, V.A.; Sharnin, V.A. *Rus. J. Phys. Chem.,* A. 2010, *vol* 84, 329-332.
[65] Usacheva, T.R.; Kuz'mina, I.A; Dzumasheva, M.O.; Sidorenko, N.S.; Sharnin, V.A. *Izv. Vyssh. Uchebn. Zaved. Khim. Khim. Tekhnol.* 2010, vol 53, 51-54.
[66] Usacheva, T.R.; Kuz'mina, I.A.; Sharnin, V.A.; Sidorenko, N.S.; Voronina, S.I. *Rus. J. Phys. Chem.,* A. 2011, vol 85, 952-955.
[67] Smirnov, V.I.; Badelin, V.G. *Rus. J. Phys. Chem.,* 2003, vol 77, 713-716.
[68] Smirnov, V.I.; Badelin, V.G.; Megevoi, I.N. *Izv. Vyssh. Uchebn. Zaved. Khim. Khim. Tekhnol.* 2003, vol 46, 90–93.
[69] Smirnov, V.I.; Badelin, V.G. *Biophysics*, 2004, vol 49, 375-380.
[70] Usacheva, T.R.; Ledenkov, S.F.; Sharnin, V.A.; Gzheidzyak, A. *Izv. Vyssh. Uchebn. Zaved. Khim. Khim. Tekhnol.* 2000, vol 43, 87-89.

In: Glycine
Editor: Wilhelm Vojak

ISBN: 978-1-62417-263-2

Chapter 2

One-Pot Electrosynthesis of Polypeptides on Surfaces in Concentrated Amino Acid Based Electrolytes: Study of Concentrated Glycine Based Electrolytes and Application to Solid pH Sensor

Guillaume Herlem[1,*], Hatem Boulhadour[1], Alexandros Antoniou[2], Freddy Torrealba Anzola[3] and Tijani Gharbi[1]

[1]Nanosciences Lab, Imagery and Therapeutics, UFR ST,
CHU Jean Minjoz, University of Franche-Comte, France
[2]Technological Institute of Larissa,
Department of Medical Laboratories, Greece
[3]UICM-Departamento de Física,
UCLA-Decanato de Ciencias y Tecnología, Venezuela

[*] To whom all correspondence should be addressed: Email: guillaume.herlem@univ-fcomte.fr.

Abstract

The anodic oxidation of concentrated based glycine electrolyte on smooth platinum electrode yields a strongly grafted polyglycine-like coating. This electrochemical reaction occurs in an irreversible way on platinum but also on other material such as graphite. Contrary to studies performed in diluted medium, spectroscopic surface analysis revealed the presence of peptide bonds in the resulting thin film layer on platinum with polyglycine like (type II) structure. Since the electrodeposited polyglycine like coating at alkaline pH have chemical groups with affinity towards protons, it was used as transducer and embedded in a solid potentiometric pH sensor. Local pH measurement can be achieved by using miniaturized electrodes thanks to photolithography process. The couple (silver chloride as reference electrode, polyglycine based platinum electrode as working electrode) of microelectrodes gives linear potentiometric responses versus pH in the range [2 - 12], reversibly and with a sensitivity of 52.4 mV/pH (for 1 mm electrode diameter). This solid pH sensor was tested over a period of thirty days and its potentiometric behavior was compared to other polymer based pH electrodes such as linear polyethylenimine (L-PEI) and polyaniline (PANI). Quality control tests can be considered by combining miniaturized pH sensors with other measurements.

Keywords: Glycine, polyglycine, pH sensor, electropolymerization, photolithography, XPS

1. Introduction

A few years ago, the anodic oxidation of pure ethylenediamine (EDA) leading to L-PEI thin film coatings on surface electrodes has been demonstrated [1,2]. From previous results, we have underlined that diamines such as EDA or 1,3-diaminopropane (1,3-DAP) must be primary for electrodeposition to occur and that the vicinal carbon not have to be substituted [3]. From this finding, the aim of this work was to continue our exploration of the anodic oxidation of bifunctional molecules including one primary amine group without substituted α carbon in their structure. In this connection, glycine was a candidate of choice even if it was the subject of many works for long. The primary interest of glycine lies in the fact that it is the simplest amino acid and is the only one that is not optically active (it has no stereoisomer). It is essential for the biosynthesis of nucleic acids as well as porphyrins, creatine phosphate, and other amino acids. Glycine, first isolated in 1820 from gelatin, is also similar to γ-aminobutyric acid and glutamic acid in the ability to inhibit neurotransmitter signals in the central nervous system. Since no toxicity has emerged from glycine studies, this molecule is used in a wide range of applications such as flavor enhancers and maskers, pH buffers and stabilizers, ingredients in pharmaceutical products, and as a chemical intermediate.

The simplest synthetic polypeptide which comes from exists in two different crystalline forms; polyglycine I (PGI), a planar (β) zigzag conformation and polyglycine type II (PGII), a helical conformation. As evocated previously, PG is also involved in many applications such as linkers in molecular biological techniques and biodegradable nylon polymers. It was shown recently that this polymer can form stable non covalent structures by self-assembly through intermolecular hydrogen bonding [4,5].

The first electrochemical studies of amino acids date back to 1968 and focused on their behavior in strong acid solutions where amino acids leads to carboxylic acids, aldehydes, aammonia and carbon dioxide [6]. Electrochemistry of amino acids spread out until a renewal of interest at the end of the 90s for studying the adsorption, reactivity and protein-surface interactions.

From the literature, it appears that glycine due to its small size and chemical and biological importance, has attracted extensive attention in recent years with investigations on single crystal electrodes as well as on polycrystalline electrodes [7,8,9,10,11]. They consist in voltammetry studies on single crystal electrode coupled with in situ infrared (FTIRS) or Raman (SERS) spectroscopy and quartz crystal microbalance. For monocrystalline electrode [12]:

- on Pt(111) electrode, strongly bonded cyanide is formed above 0.5 V/RHE with CO_2 generation from the carboxyl group oxidation (Kolbe reaction) [13]. In addition, reversible adsorption of glycinate anions has been detected at potentials higher than 0.3 V.
- on Au(111) surface in alkaline solution there is a dissociative adsorption of glycine leading to adsorbed cyanide from -0.6 to 0.0 V/SCE to give aurous cyanide at 0.3 V.5 Above 0.3 V several reactions may occur: (1) direct oxidation of bulk glycine, (2) consumption of HO^- from H_2O hydrolysis for carbonate and bicarbonate formation and (3) gold atom dissolution from electrode surface.

For other material electrode there is quite a different result from some other works undertaken in the same period as those evocated previously. For instance, the anodic oxidation of glycine on glassy carbon electrodes (GCE) was carried out for elaborating amperometric sensors. The first application concerned the use of modified GCE from anodic oxidation of glycine 10^{-2} M dissolved in phosphate buffered solution of pH 7 for monitoring uric acid concentration [14,15,16]. The authors stated without obvious proof, e.g. spectroscopic characterizations, that after their modification procedure (four cyclic scans from -0.5 to 1.8 V/SCE), the GCE was coated by polyglycine.

Curiously, there is no data concerning highly concentrated glycine based electrolytes and their electrochemical behavior on smooth platinum electrode.

PG is synthesized industrially from years by organic route, but electrochemistry can be involved during this process has it was demonstrated for the production of α-methyl-β-(3,4-dihydrophenyl) alanine [17].

From recent works in concentrated glycine solutions, we have shown that thin film coatings can be electrodeposited [18]. The idea behind this high concentration was to enhance interaction between the electrode surface and the oxidation products of glycine. The other aspect was also to avoid the presence of any stranger species from buffer whose concentration is at least ten times that of glycine. Electrochemistry in aqueous electrolyte was performed due to the very weak solubility of glycine in nonaqueous solvents.

Consequently, the electrodeposition process from the anodic oxidation of concentrated glycine base electrolytes will be examined in the next part since covalent grafting on electrode surface occurs. In addition, the resulting polyglycine like thin layer offers affinity with protons making this coating a good candidate for pH transducer. Then, a whole solid and fast pH sensor based on PG electrodeposition in potentiometric mode will be presented with extension to quality control of cornea transplant [19].

Currently, pH is monitored in this medium by a colorimetric test (trypan blue as indicator dye) which is quite inaccurate and toxic. Usually at 8, the pH value of this culture cell decreases under bacterial infection explaining why reliable pH monitoring is the crux of culture success and why pH measurements are performed in the pH range [2 - 11].

Moreover, agribusiness and chemistry are pH electrode consuming industries and miniaturized solid state pH sensor can bring new solutions. In addition, the influence of the electrode size from millimeter to micrometer range on the response of the pH sensor is examined. PG behavior as pH sensor is also compared to other electrosynthesized polymers such as L-PEI and PANI.

2. EXPERIMENTAL

All products were from Sigma-Aldrich (France) and ultra pure water (Milli-Q, Millipore) was used when needed. The electrochemical cell was a classical three-electrode setup using an AUTOLAB PGSTAT 20 potentiostat - galvanostat (Ecochemie, The Netherlands). The reference electrode was a silver wire (SRE) and all the experiments were carried out at room temperature (293 K).

Quartz crystal microbalance (QCM) measurements were coupled to cyclic voltammetry experiments. 5 MHz polished platinum coated quartz (Maxtek, USA) used (in shear mode, AT cut) and mounted on a probe for electrochemical quartz crystal microbalance (EQCM) experiments played the role of the working electrode.

All platinum electrodes were sonicated for 30 s in water, rinsed with water and ethanol, and dried with stream of highly purified Ar immediately before use. The electrodes were then electromodified by cyclic voltammetry in the range 1.4 -1.9 V at 20 $mV.s^{-1}$ for 10 scans in 0.5 $mol.L^{-1}$ glycine solution in water charged with $NaBF_4$ 2 $g.L^{-1}$ (working electrode: Pt, counter-electrode: Pt and reference electrode: SRE, scan rate: 20 mVs^{-1}). NaOH was added to reach pH = 13.

Then, the modified electrodes were rinsed with ethanol and water and sonicated for 1 min in water to remove any physisorbed, unreacted materials from the electrode surface. The glycine-modified Pt electrode (Pt/glycine) was then placed in oven at 303 K for 4 hours before being analyzed by spectroscopy IR.

Electrode potentials in the pH buffered solutions were measured via a high impedance (> 10 MΩ) voltmeter/pHmeter (Mettler) with ± 1 mV precision.

The anodic oxidation of pure EDA electrolyte (charged with $NaBF_4$ 10^{-3} M) yields L-PEI thin film on smooth platinum electrode. This coating acted as pH transducer in previous works and was used again in this study taking into account the new electrode dimensions.

The well-known PANI is also electrodeposited from the anodic oxidation of aniline 0.2 M on platinum from neutral aqueous electrolytes charged with $LiClO_4$ 2 M [20].

Prior to pH measurements, all the modified Pt electrodes were conditioned in 0.1M KCl solutions and kept in this electrolyte when unused.

Atomic force microscopy (AFM) measurements were performed using a MultiMode AFM microscope controlled by the PicoSPM from Molecular Imaging system (Agilent now) operating in the tapping mode at a frequency of

90 kHz (Al coated Si_3N_4 tip). The polymer-modified Pt electrode (Pt/glycine, Pt/L-PEI and Pt/PANI) were then placed in oven at 303 K for 4 hours before being analysed by AFM. Ellipsometric measurements were performed via an UVISEL (Horiba, France) phase-modulated spectroscopic ellipsometer in a clean room. The incident angle was set at 70° and the light spot was 1mm in diameter. The measurements were performed in the spectral range 250-850nm, with sampling steps of 5nm.

FT-IR spectra were acquired from a Nexus 470 spectrometer (ThermoElectron, France) with the Omnisampler ATR (Attenuated Total Reflection) module equipped with Ge crystal. The coated sample was pressed on the Ge crystal under about 30,000 psi and scanned at 4 cm^{-1} resolution, from 4000 to 740 cm^{-1}, 1000 times.

XPS experiments were carried out using a VG Escalab220i XL X-ray photoelectron spectrometer. It comprised of a rotating anode X-ray source (Al Kα), a quartz crystal monochromator, a high transmission/imaging lens, a hemispherical analyser and a multi channel detector. The vacuum in the analysis chamber was approximately 10^{-10} mbar. The slit width was 0.8 mm and the take off angle was 90°.

The *ab initio* periodic calculation approach used in this work is based on the CRYSTAL06 code in which crystalline orbitals are obtained from a linear combination of atomic orbitals (LCAO) [21]. All our calculations were carried out at the B3LYP/6-31G(d) level of theory [22].

The extension of the basis set up to 6-311++G(2d,2p) level and the use of different density functionals does not change significantly the results obtained at the 6-31G(d) level [23].

3. Results and Discussion

3.1. Highly Concentrated Glycine Electrolytes

Prior to electrochemical studies, the solubility of glycine was tested in various solvents of interest especially in electrochemistry where it is indeed possible to oxidize anodically amino group on smooth platinum electrode before the solvent oxidation. From previous work, we have shown the unexpected electrochemical behaviour of primary diamines such as EDA.

In this case, EDA acted both as the electroactive species and the solvent. Here, glycine has to be dissolved. In this purpose, solubility in several solvents was examined and the results are gathered in Table 1.

Table 1. Relaxed structure of crystalline PGII at B3LYP/6-31G

	x/a	y/b	z/c
O	4.039940470617E-01	1.246858249823E-01	-3.003945624409E-02
C	1.883527027192E-01	2.723424819980E-02	6.297502108968E-02
N	2.669024917231E-01	5.305403761447E-02	2.037625291854E-01
C	9.019047792152E-02	-5.810782085112E-02	3.395824097405E-01
H	-4.888312837668E-01	1.539254981451E-01	2.189255756980E-01
H	2.486254521428E-01	-8.868006385275E-02	4.153444244101E-01
H	-1.228711209908E-01	-2.972763669120E-01	3.272494961210E-01

As expected, only water yields great solubility and we fixed the glycine concentration to 0.5 M to avoid precipitation. Despite their electrochemical stability in a wide potential window, propylene carbonate and N-methyl-pyrrolidinone were discarded because of the weak solubility reached in these solvents (less than 0.1 M). $NaBF_4$ 0.01 M was used as supporting electrolyte which is a salt with high solubility and high electrochemical stability. The different pH values of the electrolyte were adjusted either with H_2SO_4 or NaOH as needed.

3.2. Electrochemical Measurements

Due to the different forms that glycine can take in solution (acidic, basic or zwitterion) where pKa values are reported on Figure 1, it was necessary to take into account the pH influence.

Therefore, cyclic voltammetry studies were carried out the at three different pH values: 1, 6 and 13. Primary amines are easily oxidized and between 0 and 2.5 V vs. SRE (Figure 2a), there is a faradaic peak corresponding to the glycine oxidation.

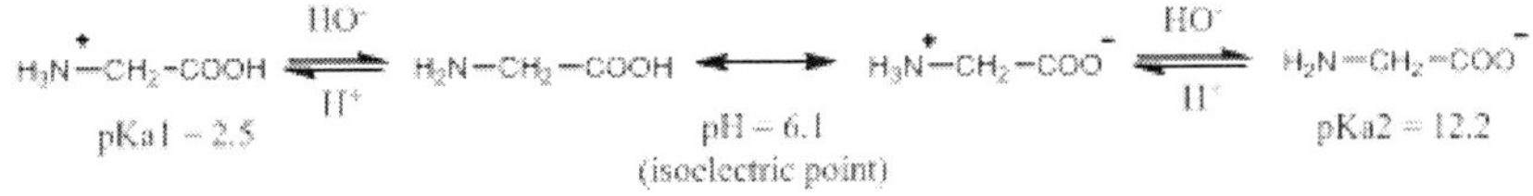

Figure 1. Forms of glycine present at various pH.

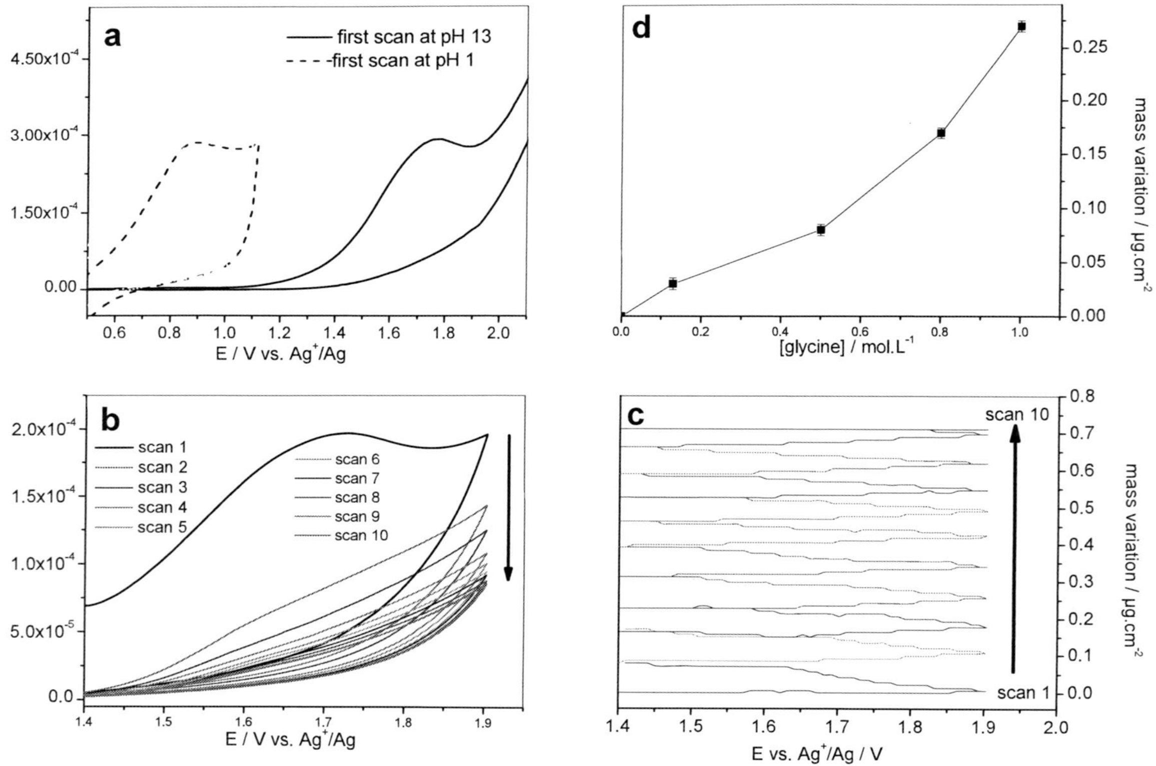
a
first scan at pH 13
first scan at pH 1
4.50x10⁻⁴
3.00x10⁻⁴
1.50x10⁻⁴
0.00
0.6
0.8
1.0
1.2
1.4
1.6
1.8
2.0
E / V vs. Ag⁺/Ag
d
0.25
0.20
0.15
0.10
0.05
0.00
mass variation / µg.cm⁻²
0.0
0.2
0.4
0.6
0.8
1.0
[glycine] / mol.L⁻¹
b
scan 1
scan 2
scan 3
scan 4
scan 5
scan 6
scan 7
scan 8
scan 9
scan 10
2.0x10⁻⁴
1.5x10⁻⁴
1.0x10⁻⁴
5.0x10⁻⁵
0.0
1.4
1.5
1.6
1.7
1.8
1.9
E / V vs. Ag⁺/Ag
c
scan 10
scan 1
0.8
0.7
0.6
0.5
0.4
0.3
0.2
0.1
0.0
mass variation / µg.cm⁻²
1.4
1.5
1.6
1.7
1.8
1.9
2.0
E vs. Ag⁺/Ag / V

Figure 2.

Its potential position depends strongly on the pH value. The higher the pH, the higher the anodic oxidation potential E. The different peak potential values found are E=0.88 V at pH=1, E=0.95 V at pH=6 and E=1.70 V at pH=13.

EQCM coupled to voltammetric measurements has proved highly effective in discriminating electrochemical phenomena at interface. The gravimetric curves obtained simultaneously correlate the mass variation at the electrode surface to the pH and the potential values. Clearly it can be observed that only at pH above 13, there is a drastic mass increase beyond the anodic peak potential at 1.7 V vs. SRE during the forward scan and the mass deposition goes on during the reverse scan.

On the contrary at pH=1, the slight mass increase occurred beyond 0.55 V vs. SRE during the forward scan is completely cancelled during the reverse scan. The electrochemical behavior of glycine at high concentration was focalized at pH 13 (above pK_a 12.2) because electrodeposition occurs and is unexpected regarding decomposition of glycine in diluted electrolytes. We performed ten scans between 1.4 and 1.9 V vs. SRE (Figure 2b). There was a current drop with decreasing peak intensity as the scans proceed. No reverse peak was observed suggesting an irreversible reaction at the platinum surface that yields an insulating coating.

Simultaneous EQCM measurements (Figure 2c) show the constant mass increase at the platinum surface by increasing the number of scans from 1 to 10 between 1.4 and 1.9 V vs. SRE. We can notice that the mass deposition is more important for the first scan than for the others.

Since electrodeposition of an insulating coating occurs at pH 13, the influence of glycine concentration on the mass electrodeposited at the electrode surface was examined. Electrochemical measurements were carried out in four different glycine concentration electrolytes: 0.13 M, 0.5 M, 0.8 M and 1.0 M. Graphic representation of the mass increase during the first scan as a function of glycine concentration (Figure 2d) shows that the higher the concentration, the higher the electrodeposited mass quantity. On the other hand, the mass increase is kept constant during the next scan for 0.13 M as for 1.0 M solution.Note that before and after the electrochemical experiments no pH change in the electrolyte solution was detected.At this stage, the electrochemical behaviour of glycine that is quite dependent upon pH solution and is consistent with similar study described elsewhere [24].

More, under alkaline conditions, some authors estimated that the strong chemisorption is observed through the unprotonated amino group [25]. In the light of spectroelectrochemisty experiments in alkaline solution, the strong chemisorption was reported to take place via the carboxylate group [26].

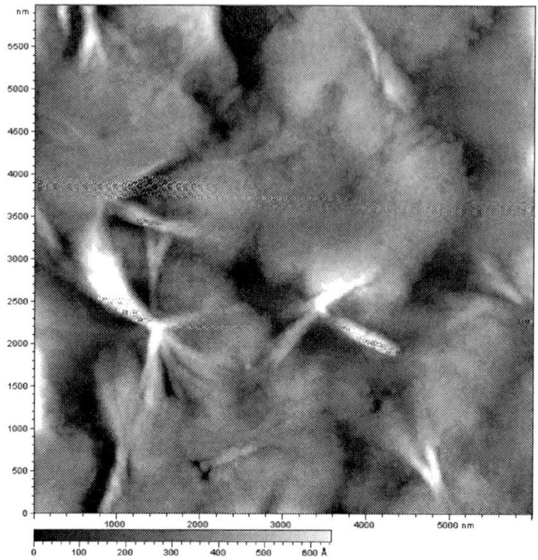

(a)

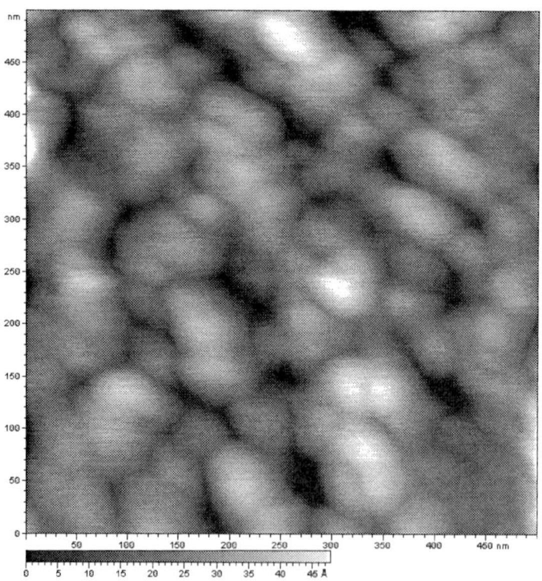

(b)

Figure 3. Contact mode AFM topography of the platinum electrode after 20 voltammetric sweeps (a) compared to (b) the bare platinum surface.

3.3. Surface Characterizations

3.3.1. AFM Characterization

Non-invasive AFM observations of electrodeposited glycine coatings were performed at air on smooth platinum coated quartz crystal used as working electrode during EQCM measurements. After ten scans, the electrode surface was rinsed with water, sonicated during 30 s and dried at 300 K. A slight milky-white complexion layer can be observed with the naked eye in light reflection on the modified surface.

Topographic image Figure 3a shows a drastic change compared to the bare platinum surface Figure 3b. The typical platinum nodules (about 50 nm diameter) have completely disappeared under the electrodeposited coating.

Contrary to the work described in the literature at diluted concentration, we are not in presence of adsorbed specie monolayer. More, the scare lines on the AFM images are characteristic of stick – slipping interactions between the tip and the coating which are in favour of a polymeric film electrodeposited on the electrode surface.

3.3.2. ATR-FTIR Spectroscopy

Platinum working electrode surfaces were analyzed by mean of ATR-FTIR spectroscopy when rinsed and dried after electrochemical experiments at pH 1, 6 and 13. The spectrum of the coating electrodeposited at pH 13, corresponding to the most abundant mass deposition condition, is shown Figure 4.Important information can be extracted from them: amide bands are present.

From these results we can see that the anodic oxidation of concentrated glycine based electrolyte leads to a passivated electrode surface with a polypeptide like coating. Peptide bond formations are probably electrocalysed during the anodic oxidation of primary amine in water.

Effectively, the anodic oxidation of $R\text{-}CH_2\text{-}NH_2$ in water yields aldehyde R-CHO. And the reaction between aldehyde and primary amine leads to amide.

In addition, the ATR-FTIR spectra from our coatings are different from the glycine (or glycine salt) one [27].

The spectral features of our coating displayed Figure 4 are almost identical to those of polyglycine II (PGII) oligomers [28]. Due to the tight binding of our coating with the platinum surface, some vibration modes can disappear and some others can be enhanced, e.g. the amide III mode in the region 1290 - 1240 cm^{-1} and the primary amine at 1100 cm^{-1}, respectively.

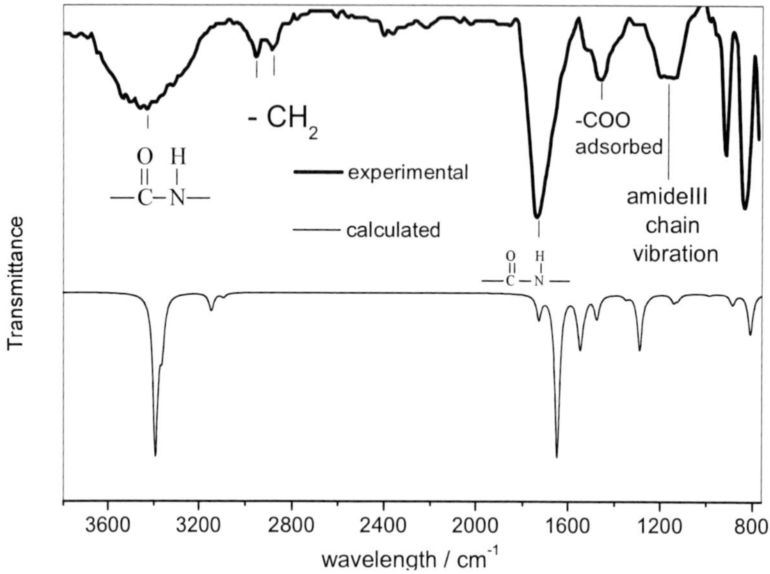

Figure 4. ATR-FTIR spectrum (experimental) of the modified platinum electrode after 20 voltammetric sweeps at pH= 13 and the calculated spectrum of bulk PG II at the B3LYP/6-31G(d) level of theory.

The presence of $-CH_2$ bending vibrations at 1450 – 1400 cm^{-1} is in favour of oligomers. But the characteristic skeletal stretching band for PGII (bulk) at 1027 cm^{-1} is not visible in our case since $-NH_2$ band is broad in this region.

3.3.3. XPS Analysis

The changes in the chemical environment of platinum surface were analyzed by XPS too. If ATR-FTIR can detect chemical groups within few micrometers, XPS can probe only depth of ten nanometers. In these conditions, XPS gives a finely detailed analysis of this coating. Figure 5 shows the XPS survey spectrum and the C 1s, N 1s and O 1s regions in the insert. The pre peak at 5 eV in the figure 5 is characteristic of a polymeric structure.

The Figure 5a in the O 1s region reveals two peaks at 531.8 eV and 536 eV. The asymmetric peak at 531.8 eV is attributed to –C=O in polyamide bond and the deconvoluted peak at 532.7 eV agrees well carboxylate binding energy. The peak at 536 eV remains unexplained.

Two C 1s peaks are clearly resolved Figure 5b. The peak at 285.5 eV can be attributed to $-CH_2$, while the other at 288.8 can be assigned to –C=O. The peak areas give a ratio of 1 –C=O for 2 $–CH_2$. The peak at 287.3 eV seems to be intrinsic to glycine system and remains unclear [9].

On Figure 5c there is one asymmetric peak in the N 1s region. Peak deconvolution gave two different environments at 400.4 eV and 399.2 eV. The lowest energy binding corresponds to amide bond whereas the other at 400.4 eV can be attributed to -(C=O)-NH-(CO)- and related to the IR strong band absorption of C=O in -(C=O)-NH-(CH_2)- between 1670 and 1790 cm^{-1}.

The XPS spectra shown in Figure 5 are radically different from those concerning glycine adsorbed on Pt(111) [18].[18] Cyanide is not present on the modified platinum surface from concentrated glycine based anodic oxidation.

From these spectroscopic measurements, probably there is glycine chemisorption on platinum via the carboxylate group at pH=13, then the anodic oxidation of primary amine group yields aldehyde in water and its reaction with amine from adjacent glycine leading to amide bond.

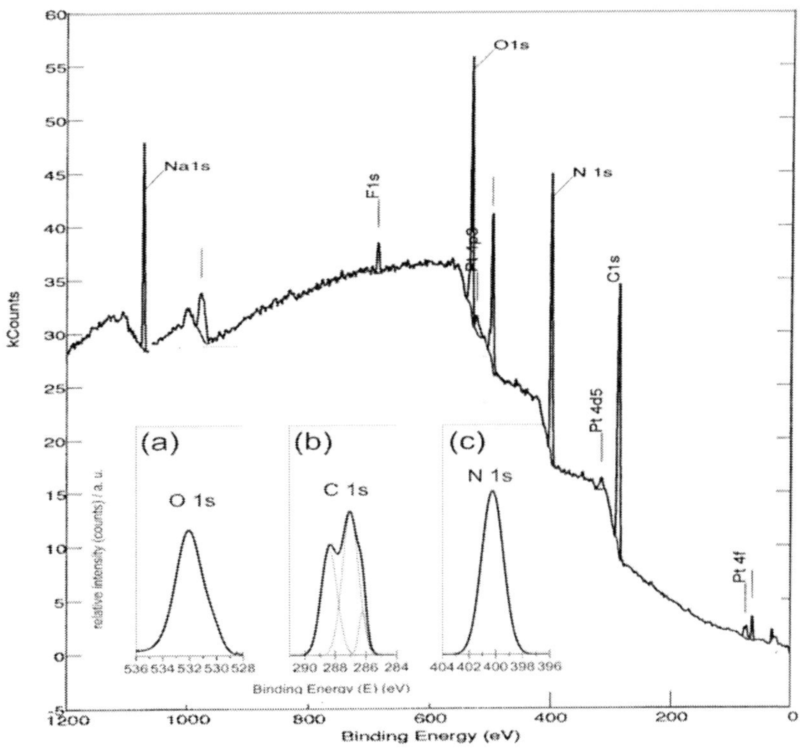

Figure 5. XPS spectrum of the coating on platinum and (a) C 1s, (b) N 1s and (c) O 1s binding energy regions.

This later step is deduced from XPS spectra and based on the peak at 400.4 eV in the N 1s region. Further reactions can lead to polypeptide chain growth on the electrode surface to give a polyglycine like coating. Although theoretical studies on PG II were carried out, no complete crystalline structure exists in the literature [30,31,32,33,34,35]. The partial existing results ($P3_1$ space group and C, N and O atomic positions) were used as the starting geometry by adding hydrogen atoms [36]. The relaxed crystalline structure at the B3LYP/6-31G(d) level of theory is reported in Table 1 (a = b = 4.8 Å, c= 9.3 Å). The structure presents the triple helix along the c axis (Figure 4) in very good agreement with the description made from X-ray powder diffraction of PG II or quantum mechanical calculations [37,38]. Concerning the XPS wide spectrum, the same profile is also observed with the calculated density of states (DOS) as shown Figure 6. The frequency calculation of the vibrational modes at the Γ point was performed from the optimized bulk lattice structure and the results are gathered in Table 2. Only three imaginary frequencies between 0 and 20 cm^{-1} corresponding to the translation modes indicate that the global minimum on the potential energy surface is reached.

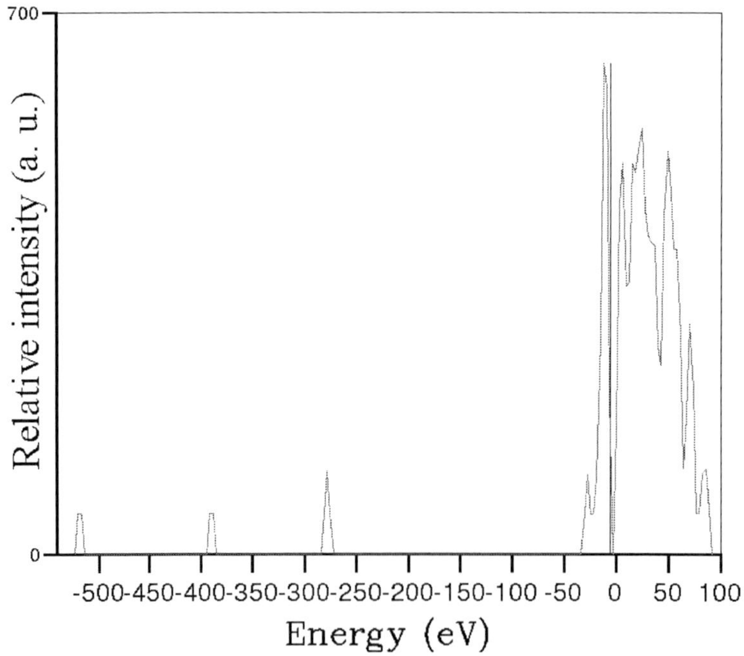

Figure 6. Calculated DOS spectrum of the crystalline PGII at B3LYP/6-31G.

Table 2. Normal modes of the crystalline PGII at B3LYP/6-31G

Modes	Frequencies (cm^{-1})	S	Intensities (km/mol)	deformation
1	-19.9313	A	(1.54)	T_z
2, 3	-4.4287	E	(0.01)	T_x, T_y
4	63.1867	A	(10.92)	ν_t(N-C-C) ⊕ ν_b(N-H)
5, 6	105.3034	E	(5.49)	ν_t(C-C) ⊕ ν_b(N-C) ⊕ ν_b(C-N-C)
7	128.161	A	(1.78)	ν_t(C-C) ⊕ ν_b(C-N-C)
8	168.0291	A	(0.02)	ν_t(C-C)
9, 10	192.1621	E	(27.77)	ν_s(O···H) ⊕ ν_t(C-C)
11, 12	272.4239	E	(14.95)	ν_t(C-N) ⊕ ν_s(O···H) ⊕ ν_b(C-N-C) ⊕ ν_b(C-C-N)
13, 14	334.3901	E	(45.75)	ν_s(O···H) ⊕ ν_b(C-C-N)
15	340.8932	A	(0.01)	ν_t(C-C) ⊕ ν_s(O···H) ⊕ ν_b(C-N-C)
16, 17	497.5683	E	(104.33)	ν_s(O···H) ⊕ ν_b(O-C-N) ⊕ ν_b(C-N-C)
18	523.9135	A	(113.66)	ν_t(C-C) ⊕ ν_s(O···H) ⊕ ν_b(C-N-C)
19, 20	565.8745	E	(86.65)	ν_t(C-C) ⊕ ν_s(O···H) ⊕ ν_b(O-C-N)
21	574.4654	A	(126.76)	ν_t(C-C) ⊕ ν_b(C=O) ⊕ ν_t(C-N)
22	644.8448	A	(45.47)	ν_s(C-C) ⊕ ν_b(C=O) ⊕ ν_b(C-C-N)
23, 24	713.0935	E	(44.75)	ν_b(C-N-C) ⊕ ν_b(C=O) ⊕ ν_b(O-C-N)
25	808.7982	A	(12.07)	ν_b(C=O) ⊕ ν_b(N-H)
26, 27	810.607	E	(602.13)	ν_b(C=O) ⊕ ν_b(N-H)
28, 29	888.2009	E	(123.89)	ν_b(C-N-C) ⊕ ν_s(C-N) ⊕ ν_s(C-C) ⊕ ν_r(CH_2)
30	894.929	A	(52.26)	ν_b(C-N-C) ⊕ ν_s(C-N) ⊕ ν_s(C-C)
31, 32	989.5726	E	(26.85)	ν_s(C-C) ⊕ ν_r(CH_2) ⊕ ν_b(N-H)
33	1036.3215	A	(0.76)	ν_b(C-N-C) ⊕ ν_b(N-H) ⊕ ν_r(CH_2) ⊕ ν_s(C-C)
34, 35	1124.4795	E	(80.44)	ν_s(C-N) ⊕ ν_s(C-C) ⊕ ν_b(N-H) ⊕ ν_{tw}(CH_2)
36	1143.2258	A	(124.74)	ν_s(C-N) ⊕ ν_s(C-C) ⊕ ν_b(N-H) ⊕ ν_{tw}(CH_2)
37	1250.7152	A	(1.35)	ν_{tw}(CH_2)
38, 39	1257.8467	E	(39.54)	ν_{tw}(CH_2)
40	1292.3091	A	(851.36)	ν_s(C-N) ⊕ ν_w(H-C-H)
41, 42	1353.2516	E	(57.64)	ν_s(C-N) ⊕ ν_s(C-C) ⊕ ν_w(H-C-H)
43, 44	1476.3936	E	(209.65)	ν_s(C-N) ⊕ ν_s(C-C) ⊕ ν_w(H-C-H)
45	1479.0858	A	(152.40)	ν_s(C-N) ⊕ ν_b(N-H) ⊕ ν_w(H-C-H)
46, 47	1527.2613	E	(97.84)	ν_s(C-N) ⊕ ν_b(N-H) ⊕ ν_w(H-C-H)
48	1545.7056	A	(332.47)	ν_s(C-N) ⊕ ν_s(C-C) ⊕ ν_b(N-H) ⊕ ν_w(H-C-H)
49, 50	1548.3278	E	(431.88)	ν_s(C=O) ⊕ ν_s(C-C) ⊕ ν_s(C-N) ⊕ ν_b(C-H) ⊕ ν_b(N-H)

Table 2. (Continued)

Modes	Frequencies (cm^{-1})	S	Intensities (km/mol)	deformation
51	1562.8006	A	(68.83)	ν_s(C=O) ⊕ ν_s(C-C) ⊕ ν_s(C-N) ⊕ ν_b(C-H)
52	1647.1651	A	(2455.55)	ν_s(C=O) ⊕ ν_w(H-C-H)
53, 54	1730.4733	E	(363.62)	ν_s(C=O) ⊕ ν_s(C-N) ⊕ ν_w(H-C-H) ⊕ ν_b(C-H)
55	3092.9124	A	(36.91)	ν_{as}(C-H)
56, 57	3093.7382	E	(22.91)	ν_{ss}(C-H)
58	3144.3359	A	(8.14)	ν_{ss}(C-H)
59, 60	3145.6485	E	(256.93)	ν_{as}(N-H)
61	3365.1201	A	(697.08)	ν_{as}(N-H)
62, 63	3393.921	E	(2353.57)	ν_{ss}(N-H)

The calculated IR spectrum (plotted with full width at half maximum (20 cm^{-1}) with normalized intensities is superimposed to the experimental one (no scaling factor used) [39]. Except the ν_s(C-H) of the theoretical spectrum around 3000 cm^{-1}, the band absorption sequence of the two spectra match very well.

If it is difficult to compare the intensities because in one case we investigate the bulk crystal and in another case we study the thin film polymer chemisorbed on a platinum surface, the characteristic skeletal stretching bands for PG II (bulk) are spread over bands between 900 and 1150 cm^{-1}. The experimental spectrum shows this visible broad band at 1100 cm^{-1}.During the electrode biasing, the anodic oxidation of glycine is maintained and the polypeptide growth is in a helical manner on the platinum surface. This original work has underlined the unexpected electrochemical behavior it exists on the anodic oxidation of aqueous concentrated glycine based electrolytes yielding polyglycine-like coating on smooth platinum surface.Surface characterizations clearly show that we are not in presence of monolayer. More, the anodic oxidation of primary amine occurs via a radical cation formation explaining the homogeneity of the resulting coating. During this step, the amide bond formation is electrocatalyzed.

ATR-FTIR and XPS revealed the chemical functionalization on the platinum surface with polypeptide chain growth helically. Furthermore, smooth platinum does not seem to be the only material where this reaction occurs. Some preliminary studies indicate that electrodeposition in the same conditions as described previously is achieved on bare gold, glassy carbon and stainless steel 316L.

PG contains basic terminal amino group usually protonated, becoming a part of bridges (intramolecular interactions) that stabilize the tertiary structure of the protein. However, it also contains less basic groups such as oxygen and nitrogen atoms where protonations are not so strong. Proton transfer from the oxygen to the nitrogen of a peptide bond exhibits a high energy barrier of about 39.1 kcal mol^{-1} [40].. On the other hand, processes that involve only oxygen atoms show significantly smaller energy barriers (16.5 kcal mol-1). [41]. Protonation can also modify the conformational equilibria of PG. [42]. In water solution polyglycine forms compact, albeit disordered globules. [43,44]. Consequently, thin PG like coatings was considered as a promising proton to pH transducer and embedded in microsystems for local pH measurements.

3.4. Design and Achievement of the pH Sensor

The micro pHsensor consists in the design of two microelectrodes deposited on a glass plate (Figure 7): one platinum working electrode and one silver reference electrode. The microelectrode connexions have rectangular ends which can be plugged to a high impedance digital voltmeter.

Glass substrates were preferred to silicon ones because of their insulating property and availability for photolithography process. The pH sensor architecture was chosen for studying the effect of the geometry and to optimize the interaction between the two electrodes. A silica layer is deposited at the final step on the plate excepted on the measuring area and the two ends allowing an effective electrical insulation. Thus, only the measure areas (Pt and Ag/AgCl) are in contact with the solution to study with no electrical leakage.

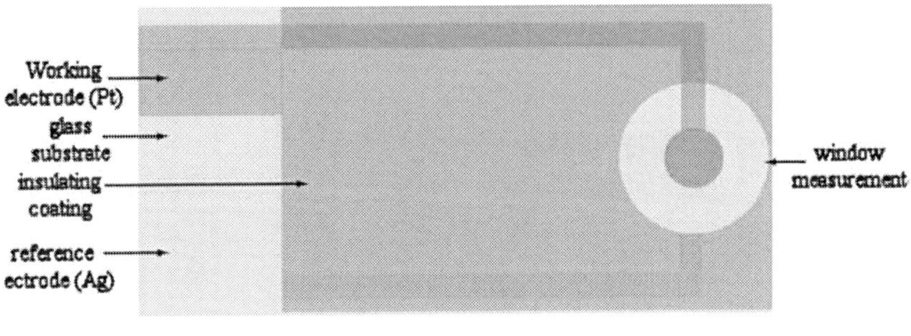

Figure 7. Schematic drawing of the micro-sensor.

Working electrode with different diameters (1000, 500, 125 and 10 μm) were fabricated in order to characterize the potentiometric response of the micro pH sensor.

3.4.1. The Photolithography Process

The first step of the microelectrode fabrication consists in drawing the mask with commercial mask design software (Cadence). Then the mask is generated by Electromask optical pattern generator.

The realization of the metallic microelectrodes is performed via a special photolithography technique called lift-off (Figure 8). First, a negative photo resist (AZ 5214 from Clariant Co.) is spread by spin coating on a 4 inch glass wafer (Figure 8.1). Then the wafer is exposed to UV radiation flux delivered by an EVG 620 aligner (Figure 8.2). Indeed, the wafer is exposed first with the mask to 40 mJ/cm² UV radiation flux, and then the wafer is exposed without mask to 210 mJ/cm² (Figure 8.3).

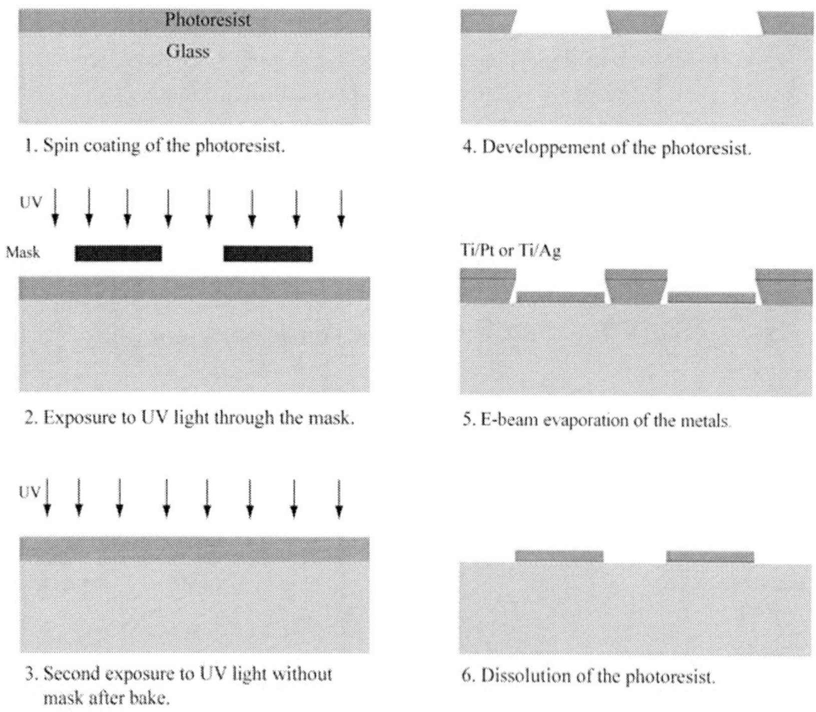

Figure 8. The lift-off process.

Figure 9. The micro-sensor realized by lift-off process.

The process is achieved by the development of the exposed photo resist (Figure 8.4). The wafer is now ready for the deposition of the metallic layers by e-beam evaporation (Figure 8.5). Photoresist is then eliminated by chemical dissolution (Figure 8.6).

After the first step of photo resist, three layers of titanium/platinum, titanium/ silver and silica have to be deposited in order to realize the micro-sensor (Figure 8.5). In this purpose, an evaporator (Alliance concept Eva 450) was used because it gives homogenous metal depositions. The wafer is introduced in the chamber which was first pumped down to 5.10-6 mbar, with both primary rotative vacuum pump and secondary turbo molecular pump.

Before the evaporation step, argon was introduced at the pressure of 3.10^{-4} mbar in an ion gun (Veeco) with two hollow cathodes in order to start argon gun and neutralizer. Etching time was 1min 30 s, just for cleaning the substrate and respects the structure of the photo resist. Titanium metal was heated by an electron beam in a graphite liner (Telemark TT10) to a speed of 5°C/s before opening the shadow in order to start the metallic layer growth. The evaporation speed was monitored in the chamber by a quartz crystal microbalance (Inficon). Titanium was just used as adhesion layer and its thickness was 20 nm. Platinum was also evaporated with the same speed but with a thickness of 150 nm.

After the lift-off of titanium and platinum and after a second step of photo resist, the wafer was introduced for the same step of cleaning and evaporation of titanium and silver. The parameter of speed and thickness were the same, it means 20 nm of titanium and 150nm of silver. By this way, the two electrodes of the sensor were realized, and the third step was a buffer layer of silica. It is necessary to use a photo resist in order to insulate the metallic arms and to let the micro sensor in contact of the solution. For good insulating properties, the layer of silica was evaporated with a high speed of 30 nm/s in order to obtain very thin grain size. The thickness of the layer was limited to 1μm because the temperature increase during the cycle and the maximum temperature that the photo resist could bear was 80°C. After the third lift-off of silica the micro sensor is ready to be functionalized electrochemically (figure 9).

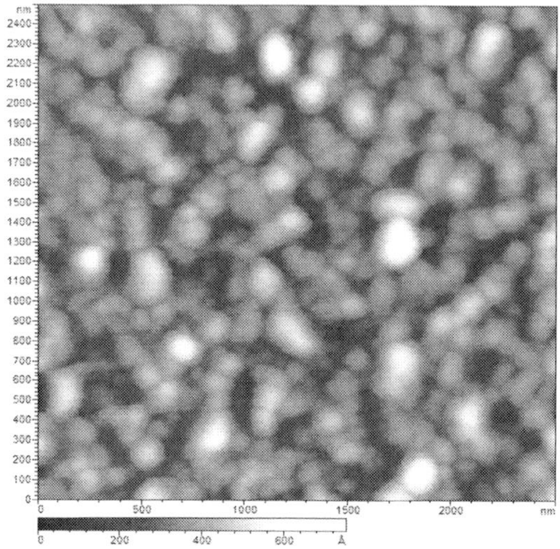

(a)

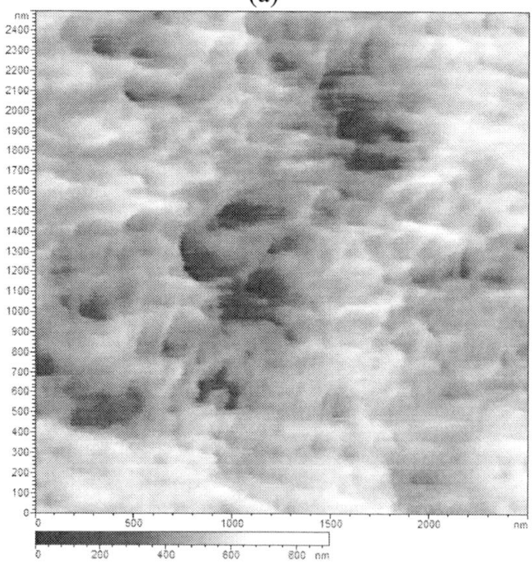

(b)

Figure 10. AFM topography (tapping mode) images of (a) the Ag reference electrode and (b) after $FeCl_3$ treatment.

3.4.2. pH Measurements

In addition to the electrodeposition of polyglycine-like coating from the anodic oxidation of concentrated glycine electrolyte, two other proton sensitive coatings were compared to the previous one and electrodeposited on smooth platinum electrode: L-PEI and PANI.

Contrary to glycine, L-PEI was electropolymerized from EDA behaves as both electrolyte and monomer to electropolymerize. EDA is aliphatic and has two primary amino groups whereas aniline is aromatic and contains a secondary amino group. Moreover, aniline has to be dissolved in a co-solvent.

Electrodeposition of the different thin polymer films (PGII, L-PEI and PANI) on smooth Pt was carried out via cyclic voltammetry method. The resulting coatings act as a transducer where the variation of the charge density occurs depending on the proton concentration. Ten scans are enough to modify irreversibly the platinum surface with the visualization of the current variation for PG and L-PEI modified electrodes whereas two scans are carried out for PANI, Figures 4b, 4c and 4d respectively.

The Ag/AgCl reference electrode used in this potentiometric pH sensor instead of silver reference electrode (SRE) avoids Ag^+ ions diffusion. Note that the potential of SRE can change with pH or the presence of Cl^- ion, but the sensor is not based on absolute potential changes referred to pH. This Ag/AgCl reference electrode is simple to prepare and reproducible. The $FeCl_3$ treatment drastically changes Ag surface (Figure 10a) where Ag smooth nodules become sharp as shown Figure 10b. [45]

The Pt/PG modified electrodes are tested in potentiometric mode as pH transducer when dipped in different buffered solutions.

At this stage, a particular care must be brought to the whole system (electrodes plus the pH meter) where stray and corrosion currents and poor electric wire shielding are a common source of errors. The electrode size effect is also taken into account (Figure 11a). In all the cases shown, there are large potential variations in the considered pH range. At the millimeter scale the potentiometric response is quasi Nernstian (52.4 mV/pH) but decreases down to 41.1 mV/pH for 10 μm electrode size which still remains satisfactory.

Concerning Pt/L-PEI electrodes, the same trends can be observed as for Pt/PG ones since L-PEI has an insulating character too but is stable in a narrower pH range [3 - 11] than that of PG (Figure 11b).Compared to Pt/PG and Pt/PEI-L, Pt/PANI (Figure 11c) modified electrodes have quasi to sub Nernstian (69.5 mV/pH) behaviors, depending on the electrode size. In fact, the potential response characterizes not only the transduction of proton concentration versus pH but also the redox sensitivity of PANI to ionic species in the buffered solutions. This chemical environment can lead to doped PANI that switches to conducting state, yielding in return side electrochemical reactions responsible for over voltage and then sub Nerstian response.

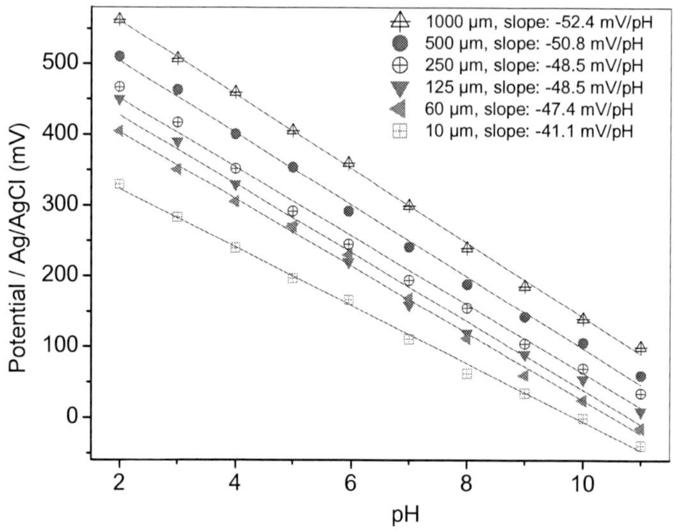

(a)

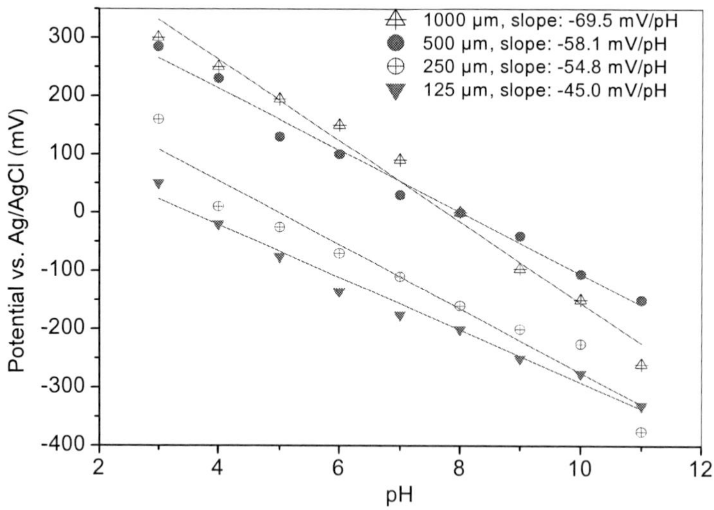

(b)

Figure 11. (Continued).

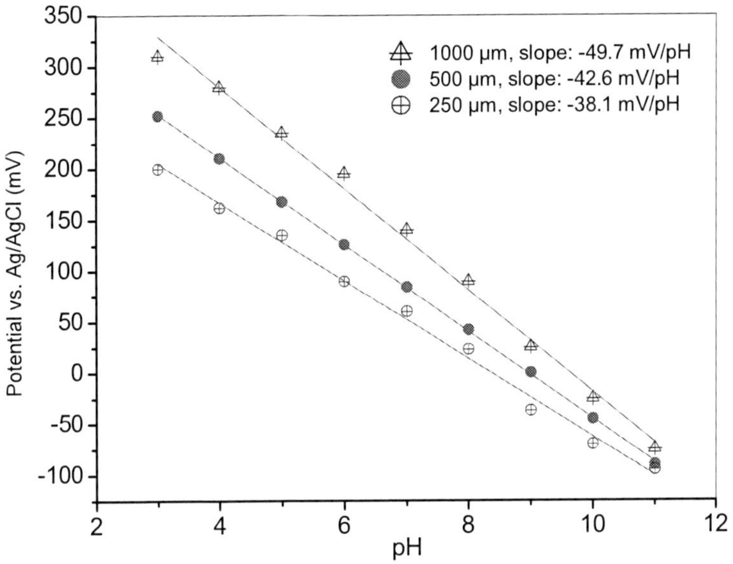

(c)

Figure 11. pH measurements on Pt electrode of different size in the pH range: (a) for Pt/PG, (b) for Pt/PEI-L and (c) for Pt/PANI.

Another drawback in using this redox polymer is its tendency to peel off in acidic medium.Other important factors to take into account are the response time of the pH measurements, the linear relationship between pH and the modified electrode potential, and the reproducibility. Concerning the reversibility of the potentiometric measurements versus pH, notice that the equilibrium potential response time decreases with the decreasing electrode size. In fact, at least two parameters are essential at this stage: the thickness of the polymer coating and the electrode area. Ellipsometry measurements have shown that after the electrodeposition process described previously, the PG coating thickness is around 15 nm. Beyond this thickness value, the response time is increased and below, the pH sensitivity is decreased. The smaller the electrode size, the smaller the sensitivity (slope). For instance, at the millimeter size, the response time is about 30 s and less than 10 s for 10 μm electrode size. The response time which is comparable to that of a glass pH electrode with millimeter size electrode (30s), is shorten drastically at the micrometer scale. The reversibility of the pH measurement is directly related to this response time. For this reason, we performed reversibility tests on Pt/PG with 10 μm size by comparing the potential responses after a pH scan from 2 to 11 and return to 2. No noticeable difference was detected. For Pt/PEI-L and Pt/PANI, the difference is barely noticeable with 10 μm electrode size too.

Globally, the potential variations versus pH of all the modified electrodes present a linear response. The linear correlation coefficients are near 1 for Pt/PG and Pt/PEI-L modified electrodes and between 0.93 and 0.98 for Pt/PANI.The Pt/PG ageing was examined by testing the responses of a newly prepared Pt/PG 60 μm size over a period of thirty days. The sensitivity of this system is slightly decreased to 42 mV/pH with a potential shift of +120 mV, which is suitable for monitoring the pH of the cornea culture bath in quality control.

CONCLUSION AND OUTLOOK

We have presented an original work based on the anodic oxidation of an aqueous concentrated glycine based electrolyte that yields polyglycine-like coating on smooth platinum electrode surface. This electrocatalyzed reaction occurs on millimetric as well as on micrometric size electrodes, leading to a solid polymeric thin film layer. Due to the presence of several chemical groups on the modified PG electrode and their affinity towards protons, the resulting coatings were used as proton transducers for solid state pH sensor. In addition to PG-like coating, different polymeric electrodeposited materials were used too such as L-PEI and PANI. From our results, PG and L-PEI coatings used as pH transducer have shown excellent linear potential responses versus pH. But polyglycine like coating gave the best sensibility and linear responses when passivating 60 μm diameter microelectrodes. Moreover, we have presented the influence of the electrode size on the pH response. If the sensitivity decreases with time, it is quite stable over a period of one month for monitoring pH variations of cornea culture bath before transplant.

PG based coating can be an alternative to colorimetric to monitoring the pH in realtime with noninvasive quantitative measurements. It is then possible to associate pH measurements to optical ones during the control quality of cornea transplant which requires passing tests with good transparency and more than 2000 endothelial cells per mm^2. In this connection, a prototype is being developed where cornea graft is lying on a glass substrate including electrode wires (Figure 12). [46] After polishing, the wire ends are flush at the glass surface and PG like coating is electrodeposited. The cornea transplant is surrounded by pH sensors and is immersed in the liquid medium coming by fluidic injection.

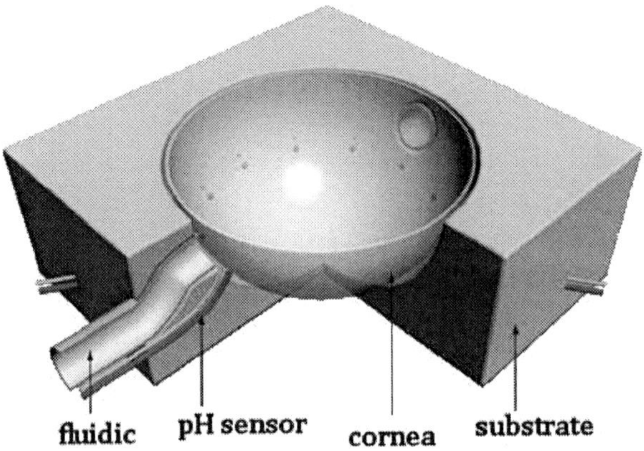

Figure 12. prototype of the device for quality control of cornea transplant.

ACKNOWLEDGMENTS

The authors gratefully acknowledge the Ecos-Nord ANUIES CONACYT program n°V08P01 (Venezuela) for financial support via scientific exchanges.

REFERENCES

[1] G. Herlem, C. Goux, B. Fahys, F. Dominati, A.M. Gonçalves, C. Mathieu, E. Sutter, A. Trokourey, J.-F. Penneau, *J. Electroanal. Chem.*, 435, 1-2, 259–265 (1997).

[2] G. Herlem, K. Reybier, A. Trokourey, B. Fahys, *J. Electrochem. Soc.*, 147, 2, 597–601 (2000).

[3] S. Lakard, G. Herlem, B Lakard, B. Fahys, *J. Mol. Struct.-Theochem,* 685, 1-3, 83–87 (2004).

[4] T. Shimizu, M. Kogiso, M. Masuda, *J. Am. Chem. Soc.,* 119, 6209 (1997).

[5] T. Shimizu, S. Ohnishi, M. Kogiso, Angew. *Chem. Int. Ed.*, 37(23), 3260 (1998).

[6] H. Lund, M. Baizer (Eds.), Organic Electrochemistry, 3rd edition, Marcel Dekker, Inc., 1991, ISBN 0-8247-8154-6, p. 601, and 3 references therein (1991).

[7] F. Huerta, E. Morallón, F. Cases, A. Rodes, J. L. Vásquez and A. Aldaz, *J. Electroanal. Chem.,* 421, 179 (1997).

[8] C.-H. Zhen, S.-G. Sun, C.-J. Fan, S.-P. Chen, B.W. Mao and Y.-P. *Chen, Electrochem. Acta.,* 49, 1249 (2004).

[9] S. M. MacDonald and S. G. Roscoe, *Electrochem. Acta.,* 42 (8), 1189 (1997).

[10] K. Ogura, M. Kobayashi, M. Nakayama and Y. Miho, *J. Electroanal. Chem.,* 449, 101 (1998).

[11] K. Ogura, M. Nakayama, K. Nakaoka and Y. Nishihata, *J. Electroanal. Chem.,* 482, 32 (2000).

[12] Lori, T. Hanawa, *Corrosion Sciences*, 43, 2111 (2001).

[13] X.-Y. Xiao, S.-G. Sun, J.-L. Yao, Q.-H.Wu, Z.-Q. Tian, *Langmuir,* 18 (2002) 6274.

[14] A.-M. Yu, H.-L. Zhang and H.-Y. Chen, *Analyst,* 122, 839 (1997).

[15] H.-Y. Chen, A.-M. Yu and H.-L. Zhang, Fresenius *J. Anal. Chem.,* 358, 863 (1997).

[16] L. Zhang, X. Lin, Fresenius *J. Anal. Chem.,* 370, 956 (2001).

[17] T. Isawaki, K. Harada, J. Chem. Soc., *Perkin Trans.* I, 1730 (1977).

[18] G. Herlem, S. Monney, V. Blondeau-Patissier, B. Fahys, T. Gharbi, *Electrochimica Acta*, 54, 27, 6797-6802 (2009).

[19] G. Herlem, R. Zeggari, J.-Y. Rauch, S. Monney, F. Torrealba Anzola, Y. Guillaume, C. André, T. Gharbi, *Talanta,* 82, 1, 417-421 (2010).

[20] J. -L. Camalet, J. -C. Lacroix, T. Dung Nguyen, S. Aeiyach, M. C. Pham, J. Petitjean, P. -C. Lacaze, *J. Electroanal. Chem.,* 485, 1, 13-20 (2000).

[21] CRYSTAL06, v1.0.2, R. Dovesi, V.R. Saunders, C. Roetti, R. Orlando, C.M. Zicovich-Wilson, F. Pascale, B. Civalleri, K. Doll, N.M. Harrison, I. J. Bush, Ph. D'Arco, M. Llunell.

[22] R. Improta, V. Barone, K. N. Kudin and G. E. Scuseria, *J. Chem. Phys.,* 114, 8, 2541 (2001).

[23] R. Improta, V. Barone, K. N. Kudin, G. E. Scuseria, *J. Am. Chem. Soc.,* 123, 3311-3322 (2001).

[24] N. C. Reynolds, B. M. Kissela, L. H. Fleminf, *Electroanalysis,* 7, 1177 (1995).

[25] G. Horányi, E. M. Rizmayer, *J. Electroanal. Chem.,* 198 (2), 393-400 (1986).

[26] S. M. MacDonald, S. G. Roscoe, *Electrochem. Acta.,* 42 (8), 1189-1200 (1997).

[27] M. Rosado, M. L. T. S. Duarte, R. Fausto, *Vibrational Spectroscopy,* 16, 35-54 (1998).

[28] K. Taga et al., *Vibrational Specroscopy,* 14,143-146 (1997).

[29] P. Löfgren, *Surface science,* 370, 277-292 (1997).

[30] V. Sasisekharan, V. N. Balaji, *Macromolecules,* 12, 28 (1979).

[31] A. M. Dwivedi and S. Krimm, *Biopolymers, Vol.* 21, 2377 (1982).

[32] R. H. Scheicher, D. Cammarere, N. Sahoo, K. Nagamine, T. P. Das, Z. *Naturforsch.,* 57 a, 523 (2002).

[33] Yun-Dong Wu, Yi-Lei Zhao, *J. Am. Chem. Soc.,* 123, 5313 (2001).

[34] S. SCheiner and C. W. Kern, *Proc. Natl. Acad. Sci.,* Vol. 75, No. 5, 2071 (1978).

[35] Y. Abe, S. Krimm, *Biopolym.,* 11, 1841 (1972).

[36] G. N. Ramachandran, V. Sasisekharan, C. Ramakrishnan, *Biochim. Biophys. Acta,* 112, 168 (1966).

[37] F. H. C. Crick, A. Rich, *Nature,* 176, 780 (1955).

[38] R. Improta, V. Barone, K. N. Kudin, G. E. Scuseria *J. Am. Chem. Soc.,* 123, 3311-3322 (2001).

[39] P. Ugliengo, D. Viterbo, G. Chiari Z. *Kristallogr..* 207, 9 (1993).

[40] Rodriquez, C. F.; Cunje, A.; Shoeib, T.; Chu, I. K.; Hopkinson A. C.; Siu K. W. M. *J. Phys. Chem.,* A, 104, 5023 (2000).

[41] Paizs, B.; Csonka, I. P.; Lendvay, G.; Suhai, S. Rapid Commun. *Mass Spectrom.,* 15, 637 (2001).

[42] P. Kulhanek, E. W. Schlag, J. Koca, *J. Am. Chem. Soc.,* 125, 13678-13679 (2003).

[43] Tran, H. T.; Mao, A.; Pappu, R. V. *J. Am. Chem. Soc.,* 130, 7380 (2008).

[44] D. Czapiewski, J. Zielkiewicz, *J. Phys. Chem.,* B, 114, 4536–4550 (2010).

[45] B.J. Polk, A. Stelzenmuller, G. Mijares, W. MacCrehanb, M. Gaitan, Sens. *Actuators* B 114, 239 (2006).

[46] G. Herlem, T. Gharbi, P. Humbert. System for detecting at least one chemical substance. US Patent *Application,* 20060121598 (2006).

In: Glycine
Editor: Wilhelm Vojak

ISBN: 978-1-62417-263-2

Chapter 3

HETEROLOGOUS EXPRESSION OF THE GLYCINE RECEPTOR CHLORIDE CHANNEL

***Qiang Shan*[1*] *and Lu Han*[2#]**

[1]Brain and Mind Research Institute, University of Sydney, Sydney, NSW, Australia

[2]Queensland Brain Institute, University of Queensland, Brisbane, QLD, Australia

ABSTRACT

The glycine receptor (GlyR) is a neuronal chloride channel, which, upon binding glycine, allows chloride anions to permeate into the neuron causing hyperpolarization of the transmembrane potential and inhibition of neuronal activity.

To investigate the structure and function of the GlyR, scientists usually isolate it by expressing it in a heterologous expression system, such as the *Xenopus laevis* oocyte or a cultured immortalized mammalian cell line, such as HEK 293.

In this chapter, protocols for the expression of the GlyR in these two expression systems are given, including preparation of recombinant cDNA, preparation of cRNA, culturing HEK 293 cells, surgical isolation

* Corresponding author: Qiang Shan. Brain and Mind Research Institute, University of Sydney, Sydney, NSW, 2050, Australia. E-mail: qshan@yahoo.com, Telephone: +61 2 9114 4032; Fax: +61 2 9114 4035.

Lu Han: Queensland Brain Institute, University of Queensland, Brisbane, QLD, 4072, Australia.

of *Xenopus laevis* oocytes, transfection of cDNA into the HEK 293 cells, injection of cRNA into the oocytes, and detection of the expressed GlyRs.

INTRODUCTION

Glycine, the simplest amino acid, is not chiral. Apart from forming peptides with the other proteinogenic amino acids, glycine also plays a prevalent physiological role in the central nervous system, where it serves as a neurotransmitter that activates the inhibitory glycine receptor (abbreviated as GlyR in this chapter).

The GlyR is widely distributed in the postsynaptic membrane of the spinal cord, brain stem and retina. When activated, the GlyR, which is a neuronal chloride channel, allows chloride anions to permeate into the neuron, causing hyperpolarization of the transmembrane potential and inhibition of neuronal activity.

The unique function of the GlyR is predominantly determined by its primary structure (i.e. its amino acid sequence). Since the GlyR was cloned, scientists have been investigating the relationship between its primary structure and its function.

Because the GlyR coexists with many other ion channels in native neuronal cells, scientists usually isolate it in a "cleaner" heterologous expression system in order to more easily investigate it. Two widely used heterologous expression systems are the cultured immortalized mammalian cell line (such as HEK 293) and the surgically-isolated *Xenopus laevis* oocyte.

In this chapter, protocols for the expression of the GlyR in both heterologous expression systems are detailed, including preparation of recombinant cDNA, preparation of cRNA, culturing HEK 293 cells, surgical isolation of *Xenopus laevis* oocytes, transfection of cDNA into the HEK 293 cells, injection of cRNA into the oocytes, and detection of the expressed GlyRs.

The heterologous system that is used in a given situation is determined by experimental requirements. Table 1 lists advantages and disadvantages of both systems.

HETEROLOGOUS EXPRESSION OF THE GLYR IN THE HEK 293 CELL LINE

Protocols

1. Preparation of Recombinant cDNA

To express the GlyR in HEK 293 cells, the cDNA that encodes the GlyR is needed. The GlyR cDNA should be subcloned into a suitable mammalian expression plasmid vector. (The authors have had success with a commercial plasmid vector, pcDNA3.1 from Invitrogen, which uses the CMV promoter to direct downstream gene expression in mammalian cells.)

Table 1.

	HEK 293 cell line	*Xenopus laevis* oocyte
Expression	-Less laborious	-More laborious
Electrophysiological detection	-Detected by patch-clamp recording -Less efficient -Intracellular solution can be adjusted	-Detected by two-electrode voltage clamp recording -More efficient -Intracellular solution cannot be adjusted
Detection by fluorescence	-Suitable for immunofluorescence confocal imaging -Not suitable for voltage-clamp fluorometry	-Not suitable for immunofluorescence confocal imaging -Suitable for voltage-clamp fluorometry

Alternatively, a stock of a plasmid vector that carries the GlyR cDNA can be requested directly from other scientists, including the corresponding author of this chapter.

Fusion of the cDNA with a gene that encodes an epitopic tag (such as FLAG) is usually required to facilitate the detection of the expressed GlyR by immunofluorescence (as described below).

After obtaining the plasmid stock, proceed by transforming it into competent *E. coli* cells, followed by minipreparation. Competent *E. coli* cells for transformation and the miniprep kit can be obtained commercially, in which case the relevant protocols from the product manuals should be followed.

A description of the processes of subcloning, gene fusion, transformation and minipreparation is beyond the scope of this chapter and can be found in textbooks on general molecular biology technique, such as *Current Protocols in Molecular Biology* [1].

2. Preparation of Cells

Materials

Ad 293 Cells

Note: The Ad 293 cell line (Stratagene) is a derivative of the HEK 293 cell line with improved cell adherence. If Ad 293 cells are used, no adherence-enhancing coating, such as poly-D-lysine, is needed.

Ca^{2+}- and Mg^{2+}-free phosphate-buffered saline (PBS) (Invitrogen)

Trypsin-EDTA solution: 0.05% (w/v) trypsin/0.02% (w/v) EDTA (Invitrogen)

Complete DMEM: Dulbecco's Modified Eagle's Medium (DMEM, Invitrogen Cat# 11995) supplemented with 10% (v/v) fetal bovine serum (FBS, Invitrogen) and 1% (v/v) penicillin-streptomycin solution (5000 units penicillin and 5000 μg streptomycin per ml, Invitrogen).

Note: Preparations of PBS, Trypsin-EDTA solution, DMEM, and penicillin-streptomycin solution are commercially available. They are likely to be of better quality than lab-prepared versions and are recommended for doing the experiments described here. The complete formulation of each solution can be found in the product information provided by its vendor.

70% (v/v) ethanol solution

75 cm^2 tissue culture flasks

35 mm tissue culture dishes

Tissue culture hood

Humidified incubator (37°C, 5% CO_2)

Light microscope

37°C water bath

Thaw the Cells

Cells are usually purchased from suppliers or requested from other scientists. They are usually stored frozen in DMEM containing the cryoprotectant DMSO (5-10%, v/v).

2.1. Add 20 ml complete DMEM to a 75 cm^2 tissue culture flask. Place it into the humidified incubator (37°C, 5% CO_2) for preconditioning for 20 mins.

2.2. Thaw the frozen cells in a cryovial in a 37°C water bath.

2.3. Take the cryovial from the water bath, wipe its external surfaces with 70% (v/v) ethanol solution, and then place it into the tissue culture hood.

2.4. Remove the tissue culture flask from the incubator. Using a pipette, gently transfer the thawed cell suspension from the cryovial to the tissue culture flask. Gently swirl the flask to distribute the cells evenly. Return the flask to the incubator.

2.5. After 5-6 hours, remove the tissue culture flask from the incubator and examine the flask under a light microscope. A majority of the cells should have adhered well to the bottom surface of the flask.

2.6. Remove the medium from the flask by aspiration, and replace with 10 ml fresh complete DMEM pre-warmed to 37°C. Return the flask to the incubator.

2.7. Using a light microscope, check the cell confluency on a daily basis. The confluency should double approximately every 24 hrs.

Passage the Cells

Cells should be passaged when the confluency is between 50-80%.

2.8. Remove the medium from the flask by aspiration.

2.9. Add 10 ml Ca^{2+}- and Mg^{2+}-free PBS to the flask and swirl it gently. Remove the PBS by aspiration.

Note: This step cannot be skipped. DMEM contains Ca^{2+} and Mg^{2+}, which would inhibit the activity of the trypsin used in the subsequent step. Ca^{2+}- and Mg^{2+}-free PBS must be used to wash off the residual DMEM.

2.10.Add 2ml Trypsin-EDTA solution to the flask. Return the flask to the incubator for 1 min.

Note: The digestion should be terminated when cells change from being flat to being round, which can be observed under a light microscope. Over-digestion by trypsin would reduce the cell viability.

2.11. Add 10 ml complete DMEM to the flask to inactivate the trypsin. Using a pipette, gently and repeatedly blow bubbles of air over the flask inside bottom surface to resuspend the cells.

Note: The added complete DMEM will inactivate the trypsin. Some scientists tend to tap the bottom of the flask or swirl the flask before adding the complete DMEM, in order to facilitate the cells detaching from the flask surface. In our experience, this does not help cell detachment.

Instead, it usually causes the cells to clump together, making it difficult to find isolated cells when performing electrophysiological recording in the subsequent steps.

2.12. Transfer 1-2 ml of the cell suspension from the flask into a fresh 75 cm^2 tissue culture flask that contains 10 ml complete DMEM. Swirl the fresh flask gently and place it into the incubator.

Note: The volume of cell suspension transferred is determined by the preferred date of the next passaging. For example, a transfer of 1 ml of cell suspension (~1:10 dilution) is usually suitable for the next passaging that occurs in 3-4 days. However, the volume of cell suspension transferred must be large enough to achieve an initial confluency of at least 5% to ensure a viable cell population.

3. Transfection

There are many ways to transfect the GlyR-expressing plasmids into the cultured cells. The simple, economic, and widely-used calcium phosphate-DNA precipitate method is described here.

Materials

Ca^{2+}- and Mg^{2+}-free phosphate-buffered saline (PBS) (Invitrogen)
Complete DMEM (As above.)
Poly-D-lysine solution, 50 μg/ml (Optional. See *Recipes for solutions*, below.)
$CaCl_2$, 0.25 M (See *Recipes for solutions* below.)
2× BES-buffered solution (2×BBS; see *Recipes for solutions* below.)
70% (v/v) ethanol solution
GlyR-expressing plasmid DNA
GFP-expressing plasmid DNA
35 mm tissue culture dishes
Glass cover slips (Menzei-Glaser, 12mm circle)
Tissue culture hood
Humidified incubators (37°C, 5% and 3% CO_2)
Light microscope
37°C water bath

Day 1

3.1. Set up as many 35 mm cell culture dishes as needed. To each dish, add 3-4 glass cover slips and 2 ml complete DMEM. Make sure that each cover slip stays firmly at the bottom of the dish and does not float.

Note: The cover slips can be used directly in most cases. However, if performing immunofluorescence, which usually requires several washing steps, the cover slips should be pre-coated with poly-D-lysine (PDL) solution. This can be achieved by soaking the cover slips in PDL solution (50 μg/ml) for 0.5-24 hrs. Wash the slips in water, twice, to remove the residual PDL.

3.2. Passage cells as described above. Transfer approximately 1×10^5 cells (2-3 drops of cell suspension by using 10 ml plastic pipette), obtained from step 2.11 above, to each dish. Gently swirl each dish and place it into the incubator (37°C and 5% CO_2) for 16-30 hrs.

Day 2

Repeat steps 3.3-3.5 for each dish.

3.3. Add 100 μl $CaCl_2$ (0.25 M) solution to a fresh 1.5 ml Eppendorf tube. Add 1 μg GlyR-expressing plasmid DNA to the tube, vortex briefly to mix and spin down quickly.

Note: To facilitate identification of the transfected cells, 0.1 μg GFP-expressing plasmid DNA is usually added together with the GlyR-expressing plasmid DNA to the tube. The amount of DNA should be adjusted if the transfection efficiency is low.

3.4. Add 100 μl 2×BBS drop by drop, vortex briefly to mix and spin down quickly. Leave the tube to incubate at room temperature for 10 mins

3.5. Transfer the solution prepared in step 3.4, drop by drop, evenly into a dish as prepared in step 3.2. Gently swirl the dish and then return it to the incubator (37°C and 3% CO_2) for 18-24 hrs.

Note: The 3% CO_2 atmosphere is used to maintain the pH of the medium at a value that allows proper formation of the calcium phosphate–DNA precipitate. Prolonged incubation would reduce the cell viability as the precipitates are harmful to the cells.

Day 3

3.6. Observe the dishes from step 3.5 under a light microscope and it should be seen that even, granular precipitates form either free in the medium or adhered to the cells.

3.7. Remove the culture medium by aspiration. Gently wash the cells twice with 2 ml PBS. Remove the PBS by aspiration. Add 2 ml fresh complete DMEM and leave the dishes in the incubator (37°C, 5% CO_2) for 24-48 hrs prior to detection.

Note: By using the calcium phosphate–DNA precipitate method, around 50% of cells can be successfully transfected.

4. Detection

Because the GlyR is a glycine-activated chloride channel, its expression in the HEK 293 cell can be detected by recording the current passing through the cell membrane upon glycine application. This is usually achieved by electrophysiological patch-clamp whole-cell recording. However, a description of this technique is beyond the scope of this chapter, and the interested reader should refer to more specific books, such as *Single-Channel Recording* [2]. Described here is the technique of visually detecting the expressed GlyRs in the cell membrane (surface expression) by immunofluoresence imaging. Detection of the expressed GlyRs within the cell as well as in the cell membrane (total expression), as a control to the surface expression, is also described.

Materials

Ca^{2+}- and Mg^{2+}-free phosphate-buffered saline (PBS) (Invitrogen)
Complete DMEM (As above.)
4% paraformaldehyde solution (See *Recipes for solutions* below.)
Triton X100
Bovine serum albumin
Mouse anti-FLAG antibody (Sigma-Aldrich Cat# F1804)
Alexa Fluor 488-labelled goat anti-mouse secondary antibody (Invitrogen)
Mounting medium (DAKO)
Glass slides
Clear nail polish
(continued from step 3.7 above)

Day 5

For surface protein labeling (S-surface labeling):

4.1S. Transfer one or two cover slips to a fresh 35 mm tissue culture dish that contains anti-FLAG antibody (1:1000) in ice-cold complete DMEM, and place the dish on ice for 30 mins.

4.2S. Remove the complete DMEM by aspiration. Wash the cells briefly using ice-cold PBS, 3 times. Remove the PBS by aspiration.

4.3S. Add 2 ml paraformaldehyde (4%) solution and incubate at room temperature for 20 min.

4.4S. Wash the cells in PBS for 5 min. Repeat three times. Remove the PBS by aspiration.

4.5S. Add 2 ml PBS/0.25% (v/v) Triton X100. Incubate at room temperature for 5 min.

4.6S. Wash the cells in PBS for 5 min. Repeat twice. Remove the PBS by aspiration.

4.7S. Add 2 ml PBS/10% (w/v) bovine serum albumin. Incubate at room temperature for 1 hr.

For total protein labeling (T-total labeling):

4.1T. Transfer one or two cover slips to a fresh 35 mm culture dish that contains PBS. Rinse briefly and then remove the PBS by aspiration. Replace with 2ml paraformaldehyde (4%) solution. Incubate at room temperature for 20 min.

4.2T. Wash the cells in PBS for 5 min. Repeat three times. Remove the PBS by aspiration.

4.3T. Add 2 ml PBS/0.25% (w/v) Triton X100. Incubate at room temperature for 5 min.

4.4T. Wash the cells in PBS for 5 min. Repeat twice. Remove the PBS by aspiration.

4.5T. Add 2 ml PBS/10% (w/v) bovine serum albumin and incubate at room temperature for 1 hr.

4.6T. Remove the solution by aspiration. Add 2 ml PBS/3% (w/v) bovine serum albumin/1:1000 anti-FLAG antibody. Incubate at room temperature for 2 hrs.

4.7T. Wash the cells in PBS for 5 min. Repeat three times.

For both surface and total protein labelings: (To follow from steps 4.7S or 4.7T.)

4.8. Remove the solution from steps 4.7S or 4.7T by aspiration. Add 2 ml PBS/3% (w/v) bovine serum albumin/1:1000 Alexa Fluor 488-labelled goat anti-mouse secondary antibody (final concentration 2 μg/ml). Incubate at room temperature in a dark place for 1 hr.

4.9. Wash the cells with PBS for 5 min. Repeat three times. Remove the PBS by aspiration.

4.10. Mount the cover slip, cell-side-down, onto a fresh glass slide by using a small drop of mounting medium. Dry the slide in a dark place. The following day, seal the edge of the cover slip to the glass slide with clear nail polish.

4.11. Proceed with confocal imaging or store the slides at 4 °C in a dark place. For confocal imaging protocols, refer to the manual of the relevant confocal microscope.

Note

1 All solutions and containers used for culturing cells must be sterile, and general sterile technique should be followed at all times when performing cell culture.

2 A common mistake when culturing cells is that the experimenter tries to remove as much of the medium or solution as possible before replacing with fresh one. Cells will die when they are left dry, even for a short time. Therefore some amount of medium or solution should be always left in the culture dish.

3 t should be noted that most commercially-available DMEMs contain a high concentration of glycine. GlyR chloride channels expressed in the cells would be continuously activated by the ambient glycine in such media, and the constant Cl^- influx would kill the cells. This is especially true for any mutant GlyR that is very sensitive to glycine. In our experience, the mutant GlyR that has a glycine EC_{50} equal to or higher than that of the WT GlyR (around 30 μM) usually survives well without any extra treatment. For the mutant GlyR with an EC_{50} between 1 and 30 μM, the channel blocker picrotoxin (final concentration 100 μM) should be added to the medium. For the more glycine-sensitive GlyR with an EC_{50} below 1 μM, the competitive antagonist strychnine (final concentration 1 μM) as well as picrotoxin (final concentration 100 μM) should be added to the medium. These extra treatments should start before transfection.

4 The transfection efficiency depends heavily on the pH of the 2×BBS, and, therefore, the pH of the 2×BBS should be precisely measured and adjusted to the optimal value.

Recipes for Solutions

2×BES-Buffered Solution (2×BBS)

50 mM N,N-bis(2-hydroxyethyl)-2-aminoethanesulfonic acid (BES)
280 mM NaCl
1.5 mM Na_2HPO_4
Adjust to pH 6.95 by using NaOH.
Sterilize by using a 0.22 μm filter.
Store in aliquots at –20°C.

Note: The pH of this solution is critical (pH 6.95 to 6.98). A tip for preparing the solution with the correct pH is to first adjust the pH roughly to somewhere in the range of 6.5-6.8.

Reserve 10% of the solution and then, finely adjust the pH of remaining solution with NaOH. If the pH happens to go beyond 6.95, use some of the reserved solution (pH 6.5-6.8) to fine tune the pH back to 6.95. Another tip is to check the pH of the new prepared solution against that of some existing stock that has been working well.

$CaCl_2$, 0.25 M

Sterilize using a 0.22 μm filter.

Store at –20°C.

4% Paraformaldehyde Solution

(Note: This must be prepared in a fume hood.)

While stirring, heat 50 ml PBS to 65°C.

While stirring, add 2g paraformaldehyde powder.

While stirring, slowly add 0.05-0.1ml 10M NaOH until the solution is clear. (The NaOH turns the paraformaldehyde into formaldehyde.)

Let the solution cool down to room temperature, and then filter though Whatman No 1 filter paper.

Store at 4°C for up to two weeks.

Poly-D-Lysine Solution, 50 μg/Ml

Sterilize by using a 0.22 μm filter.

Store at –20°C.

HETEROLOGOUS EXPRESSION OF THE GLYR IN THE *XENOPUS LAEVIS* OOCYTE

Protocols

1. Preparation of Recombinant cDNA

To express the GlyR in the *Xenopus laevis* oocyte, cRNA that encodes the GlyR is needed. The cRNA can be obtained by *in vitro* transcription of GlyR-expressing cDNA that is carried in a suitable plasmid vector. Liman *et al.* [3] have optimized such a vector, pGEMHE, which is widely used all over the world. In this vector, the T7 promoter is inserted upstream (towards the 5' end) from the multiple cloning site, where the target cDNA, wild-type or mutant GlyR cDNA in this case, is accommodated. Therefore the T7 RNA polymerase is required for *in vitro* transcription. In addition, the multiple cloning site is immediately flanked by the 5' and 3' untranslated regions of the *Xenopus laevis* beta-globin gene, in order to enhance the translation efficiency. Three restriction sites for PstI, SphI and NheI are located about 200 bases downstream (towards the 3' end) from the multiple cloning site. The restriction sites are designed to linearize the plasmid template (see below).

Protocols for subcloning cDNA into a plasmid vector can be found in textbooks on general molecular biology technique [1]. An alternative way to obtain a stock of a plasmid vector that carries the GlyR cDNA is to request it directly from other scientists including the corresponding author of this chapter. Note that for voltage-clamp fluorometry, a cysteine residue at a specific position must be introduced into the GlyR for binding sulfhydryl-specific fluorescent dyes, which can be achieved by site-directed mutagenesis [1].

After subcloning into suitable plasmids, or obtaining the GlyR-expressing plasmids from another scientist, amplify the stock by transformation of competent *E. coli* cells, followed by minipreparation, as described in the previous protocol.

Preparation of cRNA

Materials

QIAquick PCR-purification kit (Qiagen)
T7 mMessage mMachine kit (Ambion)
RNeasy mini kit (Qiagen)
RNase Away reagent (Invitrogen)
NheI, PstI or SphI restriction enzyme (NEB)
PCR machine
Centrifuge

Note: All solutions coming into contact with the RNA must be RNase-free and all equipment coming into contact with RNA should be first treated with *RNase Away* reagent.

2.1. Digest 10 μg pGEMHE plasmids that carry the GlyR cDNA by using any of the restriction enzymes, NheI, PstI, or SphI, for 2 hrs in a 100 μl volume.

Note: This process linearizes the supercoiled cDNA template in preparation for subsequent *in vitro* transcription. Any of the restriction enzymes can be chosen as long as the recognition site of the enzyme does not exist within the target cDNA, which is wild-type or mutant GlyR cDNA in this case.

2.2. Purify the digestion product using the *QIAquick* PCR-purification kit according to the product manual. The final purification product should be dissolved in 30 μl RNase-free water.

2.3. Synthesize the cRNA by using the *T7 mMessage mMachine* kit according to the product manual. Usually 1 μg linear template cDNA obtained from step 2.2 is used for a 20 μl reaction volume.

2.4. To remove the cDNA template, add 1 μl TURBO DNase (from the *T7 mMessage mMachine* kit), and mix well. Incubate at 37 °C for 15 min.

2.5. Purify the cRNA product by using the *RNeasy* mini kit according to the product manual. The final purified cRNA should be dissolved in 30-50 μl RNase-free water.

Note that because RNase is ambient, a 4 °C centrifuge is preferred and all cRNA samples should be stored on ice.

2.6. Measure the concentration of the cRNA and adjust to 200 ng/μl. Store at -70 °C.

3. Preparation of the Xenopus Laevis Oocytes

Materials

Female *Xenopus laevis* frogs (Xenopus Express, France)
0.12% (w/v) Ethyl 3-aminobenzoate methanesulfonate solution
Incubation solution (See *Recipes for solutions* below.)
70% (v/v) ethanol solution
OR-2 solution (See *Recipes for solutions* below.)
Collagenase
Surgical tools and sutures
Plastic box (with a lid) large enough to comfortably contain a frog.
Incubator (18 °C) that contains a shaker
Shaker
90 mm petri dishes
50 ml Falcon tube
Pipette pump
Paper towel

3.1. Pour 500 ml 0.12% (w/v) Ethyl 3-aminobenzoate methanesulfonate solution into a plastic box and gently lay the female frog on the solution. Cover the box with a lid. It usually takes the frog 5-10 mins to fall asleep.

3.2. Wipe the surgical tools and the bench surface where the surgery is to be performed with 70% (v/v) ethanol solution. Soak two pieces of paper towel in 0.12% (w/v) ethyl 3-aminobenzoate methanesulfonate solution and place one of them on ice. Set up a 90 mm petri dish containing 10 ml OR-2 solution.

3.3. Once the frog has fallen into sleep, collect it from the plastic box and place it on its back on the wet cold paper towel on ice. Cover the top of the torso with the other piece of paper towel.

3.4. Make a 0.5-1 cm horizontal incision on the abdomen, which is just above either leg. Then make a second incision on the peritoneum, which should expose the ovarian lobes.

3.5. Remove 2-5 ml of ovarian tissue and transfer it into the petri dish containing OR-2 solution. Suture the peritoneum and skin separately.

3.6. Wash the removed ovarian tissue 3 times with OR-2 solution. Cut the tissue into small pieces with a pair of fine scissors. Transfer the cut tissue pieces into a 50 ml Falcon tube containing 15 ml OR-2/0.15% (w/v) collagenase. The final volume including the cut tissue pieces should be around 20 ml.

3.7. Place the tube on a shaker and let it shake gently (80 rpm) for 1–2 hrs at room temperature. Check the tube every 15 min after 1 hr incubation. Immediately stop the process when the oocytes have been separated from each other.

Note: Over-digestion would cause the oocyte to become less viable and reduce the survival rate after the oocytes are injected with cRNA.

3.8. (This step is performed in parallel with step 3.7.) Rinse the ethyl 3-aminobenzoate methanesulfonate solution off the frog with water. Place the frog back into the plastic box, to which a minimal amount of clean water has been added in place of the ethyl 3-aminobenzoate methanesulfonate solution. It usually takes 10-20 mins for the frog to wake up. After it wakes, put the frog back into its holding tank. The frog can be used for a second surgery in 4-6 months.

3.9. Wash the digested oocytes with OR-2 solution 6 times over a 1 hr period, maintaining shaking at 80 rpm. Each time the solution is changed, dump any oocytes of small size that are in the upper phase. Finally, store the oocytes in incubation solution and leave them on a shaker (50 rpm) in an 18 °C incubator. The oocytes are good for injection for 2 days.

4. Injection

Materials

OR-2 solution (See *Recipes for solutions*, below.)
Incubation solution (See *Recipes for solutions,* below.)
Mineral oil
Incubator (18 °C) that contains a shaker
Nanoliter 2000 (World Precision Instruments, Inc.)
Dissection microscope
Injection dish
Glass injection pipettes (3.5" Drummond #3-000-203-G/X)
Electrode puller (Sutter)
6-well tissue culture plate

4.1. Prepare the injection needle by pulling a piece of glass injection pipette in a puller, according to the manufacturer's instructions. Break the tip of the pipette at an external diameter of 15-25 μm by using a sharp forceps (Sterilize the forceps in a Bunsen flame before use.).

Note: Too wide a tip would cause oocytes to become less viable after injection, while too thin a tip would become clogged during injection.

4.2. Transfer a number of oocytes into a 90 mm petri dish containing OR-2 solution. Separate out the healthy oocytes under a dissection microscope using a blunt glass pipette, the opening of which should be slightly wider than the diameter of the oocyte. Healthy oocytes are round, big (with a diameter of 1.2-1.5 mm), without any damage, with clear separation of dark animal and light vegetal hemispheres, and without follicular tissue attached.

4.3. Transfer the desired number of healthy oocytes into the injection dish containing OR-2 solution. (The injection dish, which is custom-made, has several grooves for holding the oocytes in place.)

4.4. Fill the injection needle with mineral oil, ensuring there are no bubbles. Mount the needle onto the Nanoliter injector. Draw 2-5 μl cRNA solution (200 ng/μl) into the tip of the needle. Inject 50 nl (10 ng) of the cRNA solution into each oocyte, at the vegetal hemisphere near the equator. A transient swelling around the injection site usually indicates a successful injection.

Note: Injecting at the vegetal hemisphere near the equator can help avoid injecting the nucleus, which is located below the animal pole. In addition, an intact animal hemisphere is desired for recording the fluorescent signal when performing voltage-clamp fluorometry, as the animal hemisphere presents a much lower background fluorescent signal than the vegetal hemisphere.

4.5. After injection, transfer the injected oocytes into 6-well tissue culture plates that contain incubation solution. Store the plates on a shaker (50 rpm) in an 18 °C incubator.

4.6. Ideally, replace the incubation solution with fresh one and dispose of any unhealthy oocytes, on a daily basis.

Note: It is necessary to allow the injected RNA time to translate completely. As in the case of HEK 293 cells, the oocytes can become unhealthy if the expressed GlyR is very sensitive to glycine or even forms a leaking channel in the absence of glycine. In such a case, 10 μM strychnine and/or 100 μM picrotoxin should be added to the incubation solution, as described in the previous section.

4.7. After 2-3 days, the expressed GlyR in the oocytes is ready for detection. The typical methods for detection are the two-electrode voltage clamp for recording the GlyR electrical channel current and voltage-clamp fluorometry for recording the GlyR fluorescent signal. Protocols for both methods can be found in more specific books [2, 4].

Recipes for Solutions

OR-2 Solution

82.5 mM NaCl
2 mM KCl
1 mM $MgCl_2$
5 mM HEPES
Adjust to pH 7.4 with NaOH.

Incubation Solution

96 mM NaCl
2 mM KCl
1 mM $MgCl_2$
1.8 mM $CaCl_2$
5 mM HEPES
Adjust to pH 7.4 with NaOH.
(The above are the ingredients for the ND96 solution.)
0.6 mM Theophylline
2.5 mM Pyruvic acid
50 μg/ml Gentamycin
5% (v/v) Horse serum

ACKNOWLEDGMENTS

We would like to thank Mr SK Brewster for helping to prepare the manuscript.

REFERENCES

[1] Ausubel, F. M. *Current Protocols in Molecular Biology*. New York: John Wiley and Sons, Inc.; 2001.

[2] Sakmann, B. Neher, E. *Single-Channel Recording* (2nd). New York: Plenum Press; 1995.

[3] Liman, E. R., Tytgat, J., Hess, P. Subunit stoichiometry of a mammalian K+ channel determined by construction of multimeric cDNAs. *Neuron*, 1992, 9, 861-871.

[4] Gandhi, C. S., Olcese, R. The Voltage-Clamp Fluorometry Technique. *Methods in Molecular Biolo*logy, 2008, 491, 213-231.

In: Glycine
Editor: Wilhelm Vojak

ISBN: 978-1-62417-263-2

Chapter 4

GLYCINE: THEORY OF THE INTERACTION WITH FAST ION RADIATION

John R. Sabin[1, 2]***, Jens Oddershede***[1, 2]
and Stephan P. A. Sauer[3]
[1]Departments of Physics and Chemistry, University of Florida, Gainesville, Florida, US
[2]Institute for Physics, Chemistry, and Pharmacology, University of Southern Denmark, Denmark
[3]Department of Chemistry, University of Copenhagen, Copenhagen, Denmark

ABSTRACT

With the advent of the use of precise ion accelerators for medical purposes, it becomes ever more important to understand the interaction of biomolecules with fast ions. Glycine is both a protein component and a model biomolecule, and is thus an important test system. In this report, we discuss various aspects of the theoretical evaluation of the properties of energy deposition associated with the collision of fast ions with glycine and its zwitterions, including differences due to molecular conformation and orientation with respect to the ion beam direction as well as due to the effect of surrounding water molecules and the state of target aggregation. Quantum mechanical calculations, which yield the dipole oscillator strength distribution of glycine are reported. The ease with which energy is absorbed from a fast ion, described by the mean excitation energy and stopping power of glycine, is reported and discussed.

I. Introduction

Glycine is perhaps, other than water, the biomolecule of the most fundamental interest to theorists: It is a model for other amino acids and thus proteins, it contains bonding structures that are typical of biochemical systems, and it is small enough that highly accurate quantum mechanical calculations of its chemical, physical, and electronic properties may be carried out.

As the use of fast ions for medical purposes, especially tumor therapy, becomes more usual, the understanding of the basic physics of the interaction of swift ions with biomaterial at clinically relevant energies (≈ 3–300 MeV/u) becomes more crucial. For example, the end result of ion radiation should be destruction of tumors without, or with only minimal, damage to surrounding tissue. Thus, understanding of the dosimetry for the interaction of fast ions with tissue becomes of primary importance for prediction of dose-depth curves and the spatial resolution of such radiotherapy beams. [1] However there are uncertainties [2] which also must be addressed. In order to understand the details of the interaction of fast ion (hadronic) radiation with biomatter, [3] one must first understand the fundamental physics of the interaction as manifest in quantities such as cross sections for fragmentation, charge exchange, ionization, *etc*. In this chapter, we discuss the interaction of fast ions with the model system, glycine, and focus on the transfer of energy from the fast ion to the molecule.

We first present the various methods used to define and calculate quantities related to energy deposition by fast ions in glycine, and then report results and the conclusions we can draw from this study.

II. Theoretical Background

1. Energy Loss Theory

When a beam of ions with charge and some kinetic energy, E, impacts matter, kinetic energy is lost from the projectile ion and transferred as kinetic and/or electronic potential energy to the target. The collision may result in various processes: the target may be fragmented, electronically or vibrationally excited, and/or undergo charge exchange with the projectile. During such processes, the energy profile of the incoming beam is shifted to lower energy and broadened, as depicted in Figure 1.

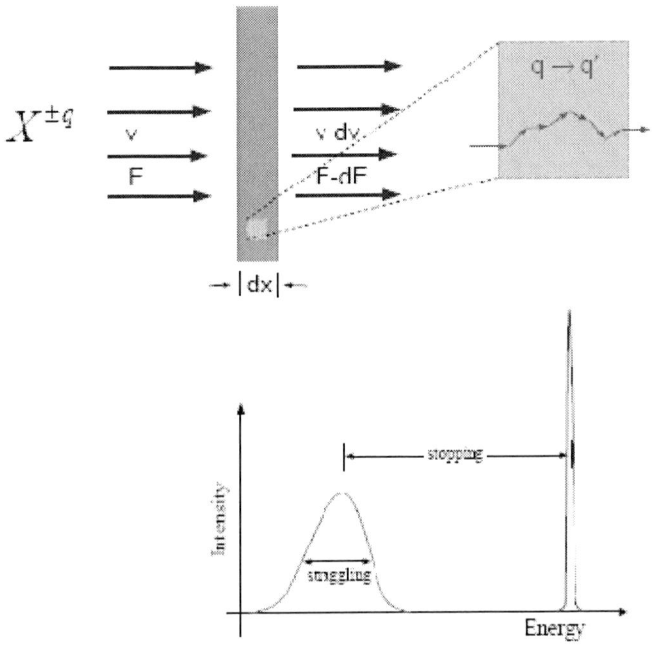

Figure 1. Schematic representation of energy loss of a fast ion to a target material.

The energy loss per unit pathlength for a fast ion in a material, known as the stopping power of the material, is generally written as a function of projectile velocity (v):

$$-\frac{dE}{dx} = nS(v) = \frac{4\pi n e^4 Z_1^2 Z_2}{m_e v^2} L(v) \tag{1}$$

Here is referred to as the stopping cross section, is the target scatterer density, Z_1 and are the projectile charge and target electron number, respectively, and and e are the electron rest mass and charge. $L(v)$, known as the target stopping number, is frequently expanded in a Born series in the projectile charge, which, for fixed projectile charge, can be written as:

$$L(v) = \sum_{i=0} Z_1^i L_i(v) \tag{2}$$

Each term can then be expanded, suppressing v, as:

$$L_i = \sum_{j=0} L_{ij} \tag{3}$$

In the following, we ignore relativistic ($\ln\frac{1}{1-\beta^2} - \beta^2$) and density ($\frac{\delta}{2}$) corrections which are small in the energy range of interest here. [4] Although there are several ways to express L, L_i, and L_{ij}, for the purposes here, the most convenient and most commonly employed are the first three terms in Eq. 2. The largest is the Bethe term L_0 : [5]

$$L_0 = \ln\frac{2m_e v^2}{I_0} - \frac{C(v)}{Z_2} \tag{4}$$

which was first introduced in this context by Bethe in 1930.[5] His original derivation considered that the projectile velocity was much larger than that of the target electrons with which it collided and exchanged energy.

As this assumption is not applicable to slow ions nor to collisions with core electrons in atoms or molecules, the shell corrections, $C(v)/Z_2$, were added to correct for this. I_0, known as the mean excitation energy, is a well defined quantity, namely the first energy weighted moment of the dipole oscillator strength distribution (DOSD) of the target system in whatever overall state the system finds itself:

$$\ln I_0 = \frac{\int \frac{df}{dE} \ln E \, dE}{\int \frac{df}{dE} dE} \tag{5}$$

The mean excitation energy is a quantity of especial interest for describing energy deposition, as it describes the ability of a target molecule to absorb energy from an ion beam (*vide infra*). Thus, were the entire set of excitation energies and oscillator strengths of the system (molecule, pure substance, or mixture, and gas, liquid, or solid) known, the unique mean excitation energy of the system would be obtainable, either theoretically or experimentally. It should be noted that, in principle, the mean excitation energy is not a parameter [6].

It is a well-defined function of the electronic excitation properties of the target and only of the target. It is not a function of projectile velocity, and, in principle, does not depend in any way on the properties of the projectile; it is a material constant of the target.

The next two terms in the Born series (Eq. 2), which are most important at low projectile velocities, include the Lindhard [7] version of the Barkas correction: [8]

$$L_1 = L_{10} = \frac{3\pi e^2 I_0}{2\hbar m_e v^3} \ln \frac{2 m_e v^2}{I_0} \tag{6}$$

and the Bloch correction: [9]

$$L_2 = L_{20} = -\sum_{l=0}^{\infty} (l+1)^{-3} \frac{v_0^2}{v^2} \cong -1.202 \frac{v_0^2}{v^2} \tag{7}$$

Here, is the Bohr velocity (2.187691 x 10^6 m/s).

In order to evaluate the leading contribution to the stopping cross section, the Bethe term, one needs to determine and thus the full dipole oscillator strength distribution for the target molecule. In this chapter we shall only deal with this lowest order approximation to hadronic stopping.

2. Electronic Structure Theory

In order to describe the energy deposition process as described above, the electronic structure of the target molecule and its excitation spectrum must first be determined, as the process primarily involves initial transfer of projectile kinetic energy to target electronic energy. The vertical electronic excitation energies, $E_n - E_0$, and associated electronic transition dipole moments for a molecule that are needed in order to obtain the mean excitation energies according to Eq. 5 can conveniently be extracted from the linear response function or polarization propagator, [10] as can be seen from its spectral representation in the basis of the eigenstates of the molecular Hamiltonian $\hat{H}_{\text{mol}}$:

$$\langle\langle \hat{P};\hat{Q} \rangle\rangle_E = \sum_{n\neq 0}\left[\frac{\langle\Psi_0|\hat{P}|\Psi_n\rangle\langle\Psi_n|\hat{Q}|\Psi_0\rangle}{E - E_n + E_0} + \frac{\langle\Psi_0|\hat{Q}|\Psi_n\rangle\langle\Psi_n|\hat{P}|\Psi_0\rangle}{E + E_n - E_0}\right] \quad (8)$$

where the sum is over all excited states $\{|\Psi_n\rangle\}$ of the system. The poles and residues of the propagator are the excitation energies and transition matrix elements, $\langle\Psi_0|\hat{P}|\Psi_n\rangle$, of the operators and , respectively. If we choose $\hat{P} = \hat{Q} = \sum_j^{Z_2} \vec{r}_j$, then the residues may be used to compute the dipole oscillator strengths of the system in the length formulation, (all quantities are in Hartree atomic units), while choices of or and lead to dipole oscillator strengths in the velocity and mixed formulations, respectively. We note that all the quantities reported in this chapter are calculated in the dipole length formulation.

In a discrete representation of the continuum, the dipole oscillator strengths of optical spectroscopy, f_{0n}, which are needed in Eq. 5, are independent of the momentum transfer q. They can then be calculated in dipole length formulation:

$$f_{0n} = \frac{2}{3}\frac{m_e}{\hbar^2}(E_n - E_0)\left|\langle\Psi_n|\sum_j^{Z_2}\vec{r}_j|\Psi_0\rangle\right|^2 \quad (9)$$

If the square of the norm is "interpreted" as the dot product of the indicated matrix element and its adjoint, then Eq. 9 represents the directionally averaged dipole oscillator strength. For the dipole oscillator strength for a particular direction, the coordinate in the expectation value should be that direction and the prefactor should be 2 instead of 2/3. The complete set of dipole oscillator strengths corresponding to excitations to all bound and continuum states of a molecule is called the dipole oscillator strength distribution (DOSD). The Thomas-Reiche-Kuhn (TRK) sum rule [11, 12, 13]

$$S_0 = \sum_{n>0} f_{0n} = Z_2 \tag{10}$$

states that the dipole oscillator strengths sum to the total number of electrons in the system. This holds not only for oscillator strengths calculated from the exact eigenstates of the molecular Hamiltonian, but also for oscillator strengths obtained with certain approximate methods, including the random phase approximation (*vide infra*), which is used here. This gives a convenient measure of how well the total DOSD is reproduced in a calculation.

In order to obtain excitation energies and transition moments from the polarization propagator one has to have another way of obtaining the propagator. This is provided by using the property that polarization propagators are the linear responses to number conserving external perturbations [10].

In practical applications, however, the exact ground state of the system is replaced by some approximate wave function which is a linear combination of antisymmetrized products of molecular orbitals, so-called Slater determinants, while one or more of the occupied molecular orbitals are replaced by virtual orbitals (excitations) in the Slater determinants or virtual orbitals by occupied orbitals (de-excitations). Approximations to the vertical electronic excitation energies are then obtained by solving the generalized eigenvalue problem and the corresponding approximations to the transition matrix elements are calculated from the eigenvectors according to Eq. 9.

This approach yields a finite number of excitations, and thus a discrete representation of the continuum. As a result, the integrations over the continuum states in Eq. 5 are done numerically using the excitation energies with energies larger than the first ionization energy of the system, called pseudo-states, as integration points. Equation 5 is then written in discrete form:

$$\ln I_0 = \frac{\sum_{n>0} f_{0n} \ln\left(E_n - E_0\right)}{\sum_{n>0} f_{0n}} \tag{11}$$

We have found that this discretization of the continuum works well provided sums over the entire excitation spectrum are taken, [14] and thus have used an explicit summation of the oscillator strengths over all bound states and the discrete continuum pseudo-states states to calculate the mean excitation energy of glycine according to Eq. 11. Experience indicates that about 12% of the mean excitation energy is due to excitation to bound states, while the remaining 88% comes from transitions into the pseudo-states for biomolecules including glycine [15]

Different types of approximate reference wave functions and corresponding sets of excitation operators can be employed in Eqs. 8 and 9. Experience also shows [16] that some amount of electron correlation is needed in order to calculate reliable spectral moments of the DOSD. One needs to calculate the propagator at least at the level of the time-dependent Hartree-Fock method, also called the random phase approximation (RPA), [17, 18, 19] which implies using a Hartree-Fock self-consistent field wave function as the function the linear response of which we are calculating.

In this approximation we are also restricting the excitation operators to the excitations from a single occupied molecular orbital to one virtual molecular orbital or vice versa. The RPA adds correlation in both ground and excited states in a balanced way. [19] Alternatively time-dependent density functional theory (TD-DFT) [20, 21] can successfully be employed in the calculation of DOSDs and thus mean excitation energies. [22, 23, 24, 25]

It should be noted that in the RPA, the dipole oscillator strengths calculated in dipole velocity, dipole length, or mixed representation and all sum rules, would be identical and the TRK sum rule, Eq. 10, would be fulfilled exactly, *i.e.* be equal to the number of electrons, if the computational basis were complete. [18, 26, 27]

Comparison of the oscillator strengths calculated in the different formulations thus gives an additional measure of the completeness of the computational basis in addition to the fulfillment of the Thomas-Reiche-Kuhn sum rule (*vide infra*).

III. RESULTS

1. Glycine structure

The first step in determination of the interaction of hadronic radiation with glycine is to determine its physical and electronic structure Glycine, H_2N-CH_2-COOH, is the simplest of the amino acids, and contains 40 electrons. Minimum energy structures of the neutral molecules were determined [15] with density functional theory, employing the B3LYP functional [28] and a standard 6-31+G(d,p) basis [29] using the Gaussian 03 program. [30] Three local minima for the isolated glycine molecule were found, and are depicted in Figure 2. It should be noted that structure (a) is the most stable and structure (b) is the least stable, but the difference among the three structures in total energy is only on the order of half a kcal/mole. [15] After the structures of interest were determined, the molecules were re-oriented in such a way that the carboxyl group was placed in the xy-plane.

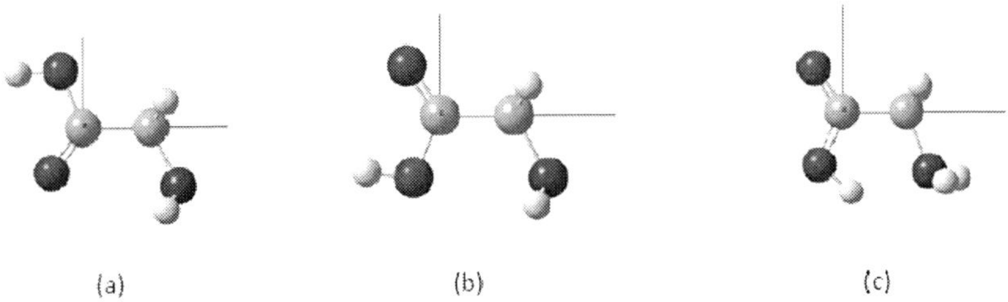

Figure 2. Optimized geometries of the three local minima of glycine. The carboxyl group lies in the xy-plane with the y-axis pointing upwards and the x-axis to the right.

The most stable structure (a), which corresponds to the experimental structure, [31] has both the C-N and O-H bonds synperiplanar to the C=O bond, so there is a bifurcated hydrogen bond from the amino group to the carbonyl oxygen, $NH_2\cdots O{=}C$. There is also a second hydrogen bond internal to the carboxyl group, from the hydroxyl group to the carbonyl oxygen, $OH\cdots O{=}C$.

In the next most stable structure (b), the O-H and C=O groups are also synperiplanar, so there is still a hydrogen bond within the carboxyl group, but the C-N bond is now antiperiplanar to the C=O bond. Thus there is only a weak bifurcated hydrogen bond between the amino and hydroxyl groups

$NH_2\cdots$O-H. The next higher energy structure (c) has both C-N and O-H bonds antiperiplanar to the C=O bond.

Similar geometrical and electronic structure calculations were also carried out for the glycine zwitterion [15]

To study the effect of peptide bonding, N-glycylglycine (diglycine) was also optimized in the same manner, however, with the non-hydrogen atoms restricted to be in the xy-plane as shown in Figure 3.

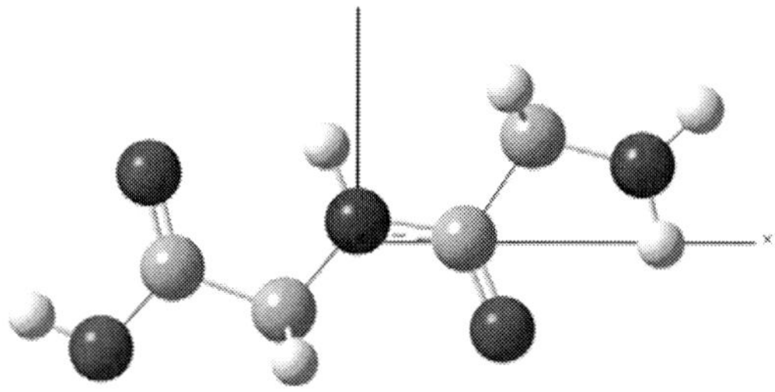

Figure 3. Optimized geometry of planar N-glycylglycine. All non-hydrogen atoms lie in the xy-plane with the y-axis pointing upwards and the x-axis to the right.

2. Glycine Mean Excitation Energies

The mean excitation energies were then determined at the RPA level using RPAC [32] and Turbomole [33, 34] and tailor-made basis sets of approximately 200 contracted functions. For details of the calculation, including basis set details, and its results, see Refs. 15 and 22. However, we should emphasize here our finding that sufficiently converged results for the mean excitation energies can only be obtained either with standard basis sets of at least quintuple zeta quality and augmented with enough polarization and diffuse functions or with specifically tailor-made basis sets of at least triple zeta quality [15]. The latter approach was chosen for glycine [15].

A fast ion can only produce electronic excitations in a target where the polarization of the excitation is perpendicular to the direction of the projectile beam. [35] Using individual Cartesian components of the dipole moment operator in the dipole oscillator strength, Eq. 9, one can obtain directional components I_0^x, I_0^y, and of the mean excitation energy from Eq. 11, which are then related to their isotropic value as $I_0 = \left(I_0^x I_0^y I_0^z\right)^{\frac{1}{3}}$

Using the directional components of the mean excitation energy, one can calculate the stopping power for a particular orientation of the target. As only excitations with polarization perpendicular to the projectile velocity can contribute to the stopping power, [36] and since we will generally place the principle plane of glycine in the xy-plane (see Figure 2), the projectiles moving perpendicular to the plane are described by and while those moving in the molecular plane are described by either and or by and I_0^z.

An anisotropy, *A*, in the energy deposition [37] can then be defined, arising from the relative orientation of the projectile direction with respect to the glycine molecular plane as:

$$A = -Z_2 \ln \frac{I_0^x I_0^y}{\left(I_0^z\right)^2} \tag{12}$$

The isotropic mean excitation energies, I_0, and anisotropies, *A*, for the three conformers of glycine are given in Table 1.

Table 1. Isotropic mean excitation energies (I_0) and anisotropies (*A*) for various conformations of glycine

	I_0 (eV)	*A*
Conformer (a)	71.10	9.21
Conformer (b)	71.08	9.23
Conformer (c)	71.03	9.41

The isotropic mean excitation energy is basically the same in all three conformations, as is also the case for the anisotropy of glycine.

Thus, as the isotropic mean excitation energy of glycine does not change significantly with respect to the molecule's conformation, then energy deposited by the beam in glycine will not depend on the molecular conformation of the three lowest lying conformers.

Calculation of the directional components of the mean excitation energy [15] for conformer (a) gives values of I_0^x, I_0^y, and I_0^z of 69.1, 67.7, and 76.8 eV, respectively. If one defines a mean excitation energy parallel to the molecular plane, $I_0^{\|} = \left(I_0^x I_0^y\right)^{1/2} = 66.3\ \text{eV}$, and one orthogonal to the plane,

$I_0^{\perp} = I_0^z = 76.8$ eV, then the mean excitation energy is largest, and thus the stopping power smallest, orthogonal to the molecular plane. Thus, as the stopping power goes as $\ln I_0$, molecular orientation does make a difference, but only on the order of 3%.

Calculations on the glycine zwitterions, [15] in either the gas phase or solvated geometry, give very similar results to the isolated glycine results, with and $I_0^{solvated} = 70.79\,eV$. The calculated anisotropies are a little smaller, [15] [15] but the conclusions are the same.

As glycine is a prototypical biomolecule, and most biology takes place in an "aqueous" environment, we consider the effect of solvation on the mean excitation energy of glycine. As glycine in solution is mostly in the zwitterionic form, the calculations were done only on the solvated glycine zwitterion. [38]

To investigate solvated glycine, we employ a polarizable quantum mechanics/molecular mechanics, QM/MM, scheme. [39] This scheme is appropriate as it takes hydrogen bonding between the solvent and the solute into explicit account. The glycine solute was treated quantum mechanically using either the RPA based scheme described above, [15] or density functional theory (DFT) with the hybrid PBEO functional. [40] Molecular dynamics were carried out using 511 water molecules in the QM/MM calculations. The RPA results for glycine give in *vacuo* and $I_0^{RPA-s} = 69.79 \pm 0.01\,eV$ when solvated. The DFT results are similar, with and $I_0^{DFT-s} = 69.28 \pm 0.01\,eV$. In all cases the anisotropies are $A \approx 3.5$. In both cases the solvent shift is negative. Thus, solvation, including hydrogen bonding, reduces the mean excitation energy of glycine and thus increases the stopping power. In both cases the effect of solvation is rather small. It should be noted, however, that the solvent shift is considerably larger than the effect due to inclusion of correlation, which is consistent with previous results. [41]

There has been some recent interest in N-glycylglycine or diglycine. [42] As it is the simplest dipeptide and thus a model for the hadronic radiation effects on protein, its mean excitation energy was also calculated at the RPA level using the same basis set and programs as for the monomer. The isotropic mean excitation energy, $I_0 = 70.68$ eV, is almost identical to the one of the glycine monomer but slightly smaller.

The calculated directional components of the mean excitation energy for N-glycylglycine are $I_0^x = 65.6$ eV, $I_0^y = 68.4$ eV, and $I_0^z = 79.4$ eV, respectively, and differ thus slightly more from the corresponding values for the glycine monomer than the isotropic value. While the component parallel to the molecular plane, $I_0^{\parallel} = \left(I_0^x I_0^y\right)^{1/2} = 67.0$ eV, is again almost equal in diglycine and glycine monomer, it is the orthogonal to the plane component, $I_0^{\perp} = I_0^z = 79.4$ eV, which is larger in diglycine leading to a smaller stopping power for ion beams orthogonal to the molecular plane than in the monomer.

3. Bragg Rule for Mean Excitation Energies

From Eq. 11, it is clear that excitations with large excitation energies that have large oscillator strengths contribute most to the mean excitation energy of a molecule.

As the perturbation on the dipole oscillator strength distribution of a collection of atoms is not large on molecule formation, one could approximate the stopping power of a molecule as a weighted sum of the stopping powers of the constituent atoms:

$$S(v)^{molecule} = \sum_{i=atoms} S_i(v) \tag{13}$$

Such a scheme is known as the Bragg Rule, [43] and can be written in terms of atomic mean excitation energies as:

$$\ln I_0^{molecule} = \frac{1}{Z_2} \sum_{i=atoms} \omega_i \ln I_0^i \tag{14}$$

Here is the number of electrons associated with atom and $Z_2 = \sum \omega_i$ is the total number of electrons in the molecule. Using standard atomic mean excitation energies, [44] this formulation gives $I_0^{atoms} = 78.15\,eV$, for glycine, some 10% higher that the full quantum mechanical calculations reported above.

More chemistry can be introduced into the formulation by calculating the contribution to the mean excitation energy due to atomic cores and to chemical bonds. For example the mean excitation energy for methane would be written as:

$$\ln I^{molecule} = \frac{1}{N}\left[\sum_i^{cores} \omega_i \ln I_i^{core} + \sum_j^{bonds} \omega_j \ln I_j^{bonds}\right] \quad (15)$$

In this scheme, methane consists of one carbon core and 4 C-H bonds. The total mean excitation energy can be calculated for simple molecules by methods described above, as can the core mean excitation energies, yielding a variety of core and bond (CB) mean excitation energies. [24, 40, 45]

This scheme, which can be extended to fragments larger than just two atom bonds, gives a mean excitation energy for glycine of $I_0^{CB} = 73.95\,eV$, [24] considerably better than the atomic Bragg Rule value (I_0^{atoms}).

Applying the fragment Bragg rule using the cores and bonds scheme to diglycine, we obtain 73.49 eV; very close to and smaller than the Bragg rule monomer value, but again slightly larger than the full quantum mechanically calculated value above. In fact, as most amino acids, and thus polypeptides have very similar compositions, one would expect most of them to have similar mean excitation energies, and thus very similar responses to hadronic radiation. This is indeed found to be the case [46]

CONCLUSION

Glycine is a remarkable molecule as it is the smallest intrinsic biomolecule. It is large enough to have biological relevance and small enough to be subject to theoretical study. Glycine is a model for other amino acids and thus polypeptides. One of the important properties of glycine, and thus, by analogy, the other amino acids and polypeptides, is its ability to absorb energy from fast ion radiation. Understanding of the details of such energy deposition in biomolecules is not only important for the developing treatment of tumors with hadronic radiation, but it has also recently been discussed [47, 48] that glycine might be the first biomolecule formed in interstellar space by molecular interaction with fast ions, and thus an early precursor to life. In any case a broad understanding of the glycine-ion interaction is important in many aspects.

It is clear that glycine is remarkably stable with respect to energy deposition by hadronic radiation. Using the mean excitation energy as a measure of the stopping power, it is evident that several factors such as molecular conformation, solvation, and peptide formation affect the mean excitation energy, and thus the stopping power, of glycine only slightly.

On the other hand, the orientation of the molecule with respect to the beam can significantly affect the mean excitation energy.

ACKNOWLEDGMENT

S.P.A.S. acknowledges support by grants from the Danish Center for Scientific Computing, the Danish Natural Science Research Council/Danish Councils for Independent Research (Grant No. 272-08-0486) and the Carlsberg foundation.

REFERENCES

[1] P. Andreo, *Phys. Med. Biol.* 54 (2009) N205.
[2] H. Paganetti, *Phys. Med. Biol.* 57 (2012) R99.
[3] C. von Sonntag, *The Chemical Basis for Radiation Biology*, Taylor and Francis, London, 1987.
[4] M.J. Berger and S.M. Seltzer, *Stopping Powers and Ranges of Electrons and Positrons,* ICRU Report #37, Bethesda, (1984).
[5] H. Bethe, *Ann. Phys.* (Leipzig) 5 (1930) 325.
[6] J. R. Sabin, J. Oddershede and S. P. A. Sauer, *Adv. Quantum Chem.* 65 (2012) in press.
[7] J. Lindhard, *Nucl. Inst. Meth.* 132 (1976) 1.
[8] W.H. Barkas, W. Birnbaum, and F.M. Smith, *Phys. Rev.* 101 (1956) 778; W.H. Barkas, J.N. Dyer, and H.H. Heckman, *Phys. Rev. Lett.* 11 (1963) 26.
[9] F. Bloch, *Ann. Phys.* (Leipzig) 16 (1933) 285.
[10] For a review of the theory and implementation of the polarization propagator method, see J. Oddershede, P. Jørgensen, and D.L. Yaeger, Compt. Phys. Rep. 2, 3 (1984); J. Oddershede, Adv. Chem. Phys. 69,

201(1987); S.P.A. Sauer, *Molecular Electromagnetism. A Computational Chemistry Approach*, Oxford University Press, 2011, chapters 2 and 10.

[11] W. Thomas, *Naturwiss.* 13, 627 (1925).

[12] W.Z. Kuhn, *Z. Phys*. 33, 408 (1925).

[13] F. Reiche, F. and W. Thomas, *Z. Phys*. 34, 510 (1925)..

[14] J. Geertsen, J. Oddershede, and J.R. Sabin, *Phys. Rev.* A 34, 1104 (1986).

[15] S.P.A. Sauer, J. Oddershede, and J.R. Sabin, *J. Phys. Chem.* A 110, 8811 (2006).

[16] J. Oddershede and J.R. Sabin, *Phys. Rev*. A 39, 5565 (1989).

[17] A.D. McLachlan and M.A. Ball, *Rev. Mod. Phys*. 36, 844 (1964).

[18] J.D. Rowe, *Rev. Mod. Phys*. 40, 153 (1968).

[19] Aa. E. Hansen and T.D. Bouman, *Mol. Phys*. 37, 1713 (1979).

[20] E. Runge and E.K.U. Gross, *Phys. Rev. Lett.* 52, 997 (1984); A. Dreuw and M. Head-Gordon, *Chem. Rev*. 105, 4009 (2005).

[21] J. Schirmer and A. Dreuw, A. *Phys. Rev*. A. 75, 022513 (2007).

[22] S. Bruun-Ghalbia, S.P.A. Sauer, J. Oddershede, and J.R. Sabin, *Eur. Phys. J.* D 60, 71 (2010).

[23] K. Aidas, J. Kongsted, J.R. Sabin, J. Oddershede, K.V. Mikkelsen, and S.P.A. Sauer, *J. Phys. Chem. Lett*. 1, 242 (2010).

[24] S. Bruun- Ghalbia, S.P.A. Sauer, J. Oddershede, and J.R. Sabin, *J. Phys. Chem*. B 114, 633 (2010).

[25] S.P.A. Sauer, J. Oddershede, and J. R. Sabin, *J. Phys. Chem*. C, 114, 20335-20341 (2010).

[26] R.A. Harris, *J. Chem. Phys*. 50, 3947 (1969).

[27] P. Jørgensen and J. Oddershede, *J. Chem Phys.* 78, 1898 (1983).

[28] C. Lee, W. Yang, and R.G. Parr, Phys. Rev. B 37, 785 (1988); A.D. Becke, *J. Chem. Phys.* 98, 5648 (1993).

[29] W.J. Here, R. Ditchfield, and J.A. Pople, J. Chem. Phys. 56, 2257 (1972); T. Clark, J. Chandrasekhar, and P.V.R. Schleyer, *J. Comput. Chem*. 4, 294 (1983).

[30] M.J. Frisch, G.W. Trucks, H.B. Schlegel, G.E. Scuseria, M.A. Robb, J.R. Cheeseman, J.A. Montgomery Jr., T. Vreven, K.N. Kudin, J.C. Burant, J.M. Millam, S.S. Iyengar, J. Tomasi, V. Barone, B. Mennucci, M. Cossi, G. Scalmani, N. Rega, G.A. Petersson, H. Nakatsuji, M. Hada, M. Ehara, K. Toyota, R. Fukuda, J. Hasegawa, M. Ishida, T. Nakajima, Y. Honda, O. Kitao, H. Nakai, M. Klene, X. Li, J.E. Knox, H.P. Hratchian, J.B. Cross, V. Bakken, C. Adamo, J. Jaramillo, R. Gomperts,

R.E. Stratmann, O. Yazyev, A.J. Austin, R. Cammi, C. Pomelli, J.W. Ochterski, P.Y. Ayala, K. Morokuma, G.A. Voth, P. Salvador, J.J. Dannenberg, V.G. Zakrzewski, S. Dapprich, A.D. Daniels, M.C. Strain, O. Farkas, D.K. Malick, A.D. Rabuck, K. Raghavachari, J.B. Foresman, J.V. Ortiz, Q. Cui, A.G. Baboul, S. Clifford, J. Cioslowski, B. B. Stefanov, G. Liu, A. Liashenko, P. Piskorz, I. Komaromi, R.L. Martin, D.J. Fox, T. Keith, M.A. Al-Laham, C.Y. Peng, A. Nanayakkara, M. Challacombe, P.M.W. Gill, B. Johnson, W. Chen, M.W. Wong, C. Gonzalez, J.A. Pople, Gaussian 03, revision C.02; Gaussian, Inc.: Wallingford, CT, 2004.

[31] P.D. Godfrey and R.D. Brown, *J. Am. Chem. Soc.* 117, 2019 (1995).

[32] T.D. Bouman and Aa. E. Hansen, *RPAC Molecular Properties Package*, Version 9.0, Copenhagen University, 1990.

[33] R. Ahlrichs, M. Bär, M. Häser, H. Horn, and C. Kölmel, Chem. Phys. Lett. 162, 165 (1989); M. Häser and R. Ahlrichs, J. Comput. Chem. 10, 104 (1989); O. Treutler and R. Ahlrichs, J. Chem. Phys. 102, 346 (1995); R. Bauernschmitt and R. Ahlrichs, Chem. Phys. Lett. 256, 454 (1996); S. Grimme, F. Furche, and R. Ahlrichs, Chem. Phys. Lett. 361, 321 (2002); F. Furche and D. Rappoport, Computational Photochemistry. In *Computational and Theoretical Chemistry*; M. Olivucci, Ed.; Elsevier: Amsterdam, 2005; Vol. 16; Chapter 3; M. Arnim, and R. Ahlrichs, *J. Chem. Phys.* 111, 9183 (1999).

[34] TURBOMOLE V6.0 2009, a development of University of Karlsruhe and Forschungszentrum Karlsruhe GmbH, 1989-2007, TURBOMOLE GmbH, since 2007; available from http://www.turbomole.com.

[35] M. Inokuti, W. Karstens, E. Shiles, and D.Y. Smith, *Mean Excitation Energy for the Stopping Power of Silicon from Oscillator-Strength Spectra*, Presented at the 2005 APS March Meeting, Los Angeles, California, March 21-25, 2005;http://meetings.aps.org/link/BAPS.2005.MAR.R1.55.

[36] H.H. Mikkelsen, J. Oddershede, J.R. Sabin, and E. Bonderup, *Nucl. Inst. Meth.* B 100, 451 (1995).

[37] S.P.A. Sauer, J.R. Sabin, and J. Oddershede, *Nucl. Inst. Meth.* B 100, 458 (1995).

[38] K. Aidas, J. Kongsted, J.R. Sabin, J. Oddershede, K.V. Mikkelsen, and S.P.A. Sauer, *J. Phys. Chem. Lett.* 1, 242 (2010).

[39] J. Kongsted, A. Osted, K.V. Mikkelsen, and O. Christiansen, *Mol. Phys.* 100, 1813 (2002).

[40] C. Adamo and V. Barone, *J. Chem. Phys.* 110, 6158 (1999).

[41] J.R. Sabin, J. Oddershede, and S.P.A. Sauer, *AIP Proc.* CP1080, 138 (2008).
[42] *Cf. e.g.* V. Feyer, O. Piekan, R. Richter, M. Coreno, K.C. Prince, and V. Carravetta, *J. Phys. Chem.* A 113, 10726 (2009).
[43] W.H. Bragg and R. Kleeman, *Phil. Mag.* 10, 318 (1905).
[44] M.J. Berger and S.M Seltzer, *Stopping Powers and Ranges of Electrons and Positrons*, 2nd ed., NBSIR 82-2550-A, Washington, 1983.
[45] J. Oddershede and J.R. Sabin, *Nucl. Inst. Meth.* B 42, 7 (1989).
[46] S.P.A. Sauer, J. Oddershede, and J.R. Sabin, Adv. *Quantum Chem.* 62, 215 (2011).
[47] R.E. Johnson, *Rev. Mod. Phys.* 68, 305 (1996).
[48] M. Lattelais, F. Pauzat, J. Plime, Y. Ellinger, and C. Ceccarelli, Astron. *Astrophys.* 532, A39 (2011).

In: Glycine
Editor: Wilhelm Vojak

ISBN: 978-1-62417-263-2

Chapter 5

ELECTROCHEMISTRY OF ALIPHATIC BIFUNCTIONAL MOLECULES: FROM POLYMERIC THIN FILM COATING TO SENSORS AND BIOSENSOR APPLICATIONS

Guillaume Herlem, Hatem Boulhadour, Alexandros Antoniou, Freddy Torrealba Anzola and Tijani Gharbi

University of Franche-Comte and CHU Jean Minjoz, France

ABSTRACT

Despite their toxicity and the rise of green chemistry, nonaqueous solvents play an important role in analytical electrochemistry. When used methodically, they can bring an elegant way to modify a substrate irreversibly in one-pot experiment. Electrochemistry in liquid ammonia and liquid amines contributes in this direction too. Since ammoniates and aminates (by analogy to the name hydrate for aqueous electrolytes) are very used in our group, I am doing electrochemistry in these media:

1 – to elucidate new (electrochemical) reactions, electropolymerization, etc.

2 – to better discriminate the analytes by use of several techniques in (bio) chemical sensors (bio) analytical

3 - to model and interpret the results by theoretical ab initio and docking.

I will explain, in what follows, the three points that summarize my research I initiated and developed in Besancon, France.

Keywords: Nonaqueous, concentrated electrolytes, electropolymerization, thin film layer, ethylenediamine, polyethylenimine, glycine, polyglycine, polyamines, 3-APTES, ethanedithiol, sensor, biosensor, cell culture, grafting, pH, microsystems, periodic ab initio calculations, DFT, XPS, IR-RAS

1. Synthesis and Characterization of New Reactions in Electrolytes

1.1. Inorganic Compounds Electrosynthesis

Electrolysis reaction in nonaqueous solvent remains largely unexplored and therefore unexploited. This is true for some metals in liquid ammonia and / or some liquid light amines at room temperature. As part of my activity on energy storage (and / or restitution) I performed electrosynthesis of original cathode materials: nitrides. They are prepared from the amide complex cations of transition metals mainly. Indeed, many of M^{x+} cations are acidic in liquid ammonia:

$$M^{x+} + 2\ NH_3 \rightarrow M(NH_2)_x\ (\textit{insoluble}) + xNH_4^+$$

The electrosynthesized amide precipitates and can be isolated by evaporation of ammonia (which can be recovered by condensation for further work). The heating under vacuum transforms amide into nitride. For example, with a trivalent cation, M^{3+}:

$$M(NH_2)_3 \rightarrow MN + 2NH_3$$

The extension to other solvents such as liquid amines can be interesting. Among the solvents related to liquid ammonia, ethylenediamine is a unique solvent which, thanks to the very high solvation cations, the low reactivity towards solvated electron (which is slightly soluble), used to synthesize stable "alcalides". This property led us to consider the preparation of nanoparticles of metals by electrolysis directly in these solvents.

1.2. Electropolymerization of Bifunctional Aliphatic Molecules

I discovered in 1996, it was possible to graft on an electrode in an electrochemical single operation (one-pot), linear polyethylenimine (L-PEI) thin film by anodic oxidation of ethylenediamine. The origin and originality of this discovery date back to studies on conductivities I performed in liquid amine-based electrolytes. I was led to study ethylenediamine (EDA observed) as a solvent to dissolve a large number of salts up to concentrations higher than 6 M. In addition to excellent maximum conductivities obtained (between 20 and 30 mS / cm at 20 °C) in this solvent, its anodic oxidation behavior caught my attention. Indeed there is an irreversible phenomenon of passivation on the first scan (in cyclic voltammetry) beyond 2.5 V / Ag with a current drop during the following scans. The quartz microbalance coupled with cyclic voltammetry experiments, allows visualizing this deposit.

I quickly realized the value of this thin layer electropolymerization in analytical chemistry since there is a great amino group density on a surface with good affinity toward protons. Another interesting feature of this reaction is that ethylenediamine acts as a solvent but also a monomer which avoids addition of a co-solvent. In the case of L-PEI, which has a high density of primary amine functions protonated, I quickly envisioned its use as a transducer to produce any solid pH sensor.

2. Extension to Other Bifunctional Aliphatic Molecules: Sensor and Biosensor Applications

From the discoveries presented above concerning polymer electrodeposition, I soon perceived their interest and their intended use as a transducer for the identification and analysis of chemicals analytes.

L-PEI can be electro-formed irreversibly on platinum, gold, glassy carbon, silicon p-type and FTO for example. This surface modification has led me to develop chemical sensors (E = f (pH)) and biochemical (glucose sensing) in potentiometric mode. The polymer attached to the surface of the electrode does not dissolve once immersed in an aqueous solution of biological interest.

Based on this electrochemical reaction of ethylenediamine electropolymerization, patents were filed with a financial support from the national agency for promotion of research and concerned the development of non-invasive glucose biosensors. This polymer is a remarkable compound, non-toxic, biocompatible, which is the "heart" of many biosensors and, more generally, many applications in biology. Before my discovery grafting thin layer PEI-L required several complex chemical operations.

2.1. Lab-on-a-Chip: the Bio-Nose

Eight years, we used electropolymerized L-PEI as a medium for cell culture because of its biocompatibility. The remarkable cell attachment and growth resulted in developing a project called "bio-nose." Under this name we came up with micro-containing neurons in the olfactory bulb of mouse attached to a thin layer of PEI-L electrodeposited on the center of the microelectrode. The activity of the cell is measured here, not by electrochemical technique, but by fluorescence of Ca^{2+} ions which pass through the pores of cell membranes following a stimulus (the odor). All integrated in a lab on a chip (LOC for Laboratory-on-a-chip).

Our aim was to provide a micro to millimeter scale device containing the different functional blocks needed for online analysis of families of odors in the liquid phase. The original device was designed as an assembly of several steps.

2.2. Olfactory Molecule Detection: The Bio-Nose 2

Given the difficulty of removing tissues from animals in the bio-nose project described above, it seemed desirable to exploit trans-membrane proteins of olfactory cells that transport of odorants: the OBP for odorant Binding Protein.

This work was made possible in collaboration with colleagues whose skills were recognized with the OBP. This led us to evolve the concept of bio-nose towards a new architecture consisting in grafting OBP on thin polymeric films such as polyethylenimine. The idea is to capture an odorant molecule into the cavity of the OBP and dislodge a molecular probe that fluoresces when it goes into solution. This allows us to quantify the amount of trapped molecules. Different coupling options were considered as the surface plasmon resonance or the use of ultra microelectrodes.

2.3. Detection of Milk Allergens

The work undertaken was to develop a biosensor for assaying the corresponding antibodies to a protein with high antigenicity such as β-casein, an immunoglobulin G (IgG). B-casein has strong allergenic, meaning it can cause allergies. These allergies occur in humans by producing antibodies Ig E. B-casein is one of the largest milk proteins and is present in very basic food, hence the importance of developing such a biosensor. The essential qualities of the biosensor had to meet:

- a high detection toward Ig G. Indeed, the amount of IgE is 10 000 times lower in blood than that of IgG (in the case of allergies, the rate of IgE is a thousand times the normal value).
- failure of the biosensor inhibition by IgG present in blood.

2.4. Packaging and Transportation of the Cells of the Retina and the Cornea

The cornea transplant yields excellent results. Before transplantation the cornea is taken (under sterile) and packed in a glass vial. The cornea is immediately put in a preserving liquid in a sterile manner, and is put in the oven of the Cornea Bank. Conservation of the organ transplants are performed in an oven at 31 ° C.

A series of checks are performed on both the safety of the graft but also on its quality. It consists of a double check:

- The first check on the macroscopic transparency of the cornea,
- The second check on the number of cells known as endothelial (located on the inner surface of the cornea) that contains the graft.

Transparency of long-term transplantation is directly related to the number of cells and their morphology. For this, endothelial cells are counted at the beginning and the end of storage. And a cornea is judged suitable for transplantation if it contains more than 2000 cells / mm ^{2}, if their morphology is regular and their mortality rate is less than 2%.

After these checks, about 55% of grafts are suitable for transplantation. When these controls mishandling can damage or contaminate the graft.

We proposed a novel system to respond to the problem of transport, control and conditioning of the graft. It's bottle conditioning to associate a microsystem that will ensure health monitoring without having to extract the graft of this set will be a "lab-on-chip".

3. Modeling and Interpretation of the Results by Theoretical Calculations

3.1. Modeling of Complex Systems. Case of NaI • 3.3 NH3

Ammoniates can deliver the most important ionic conductivities from the nonaqueous solvents. At the microscopic level, the concentrated electrolyte NaI • 3.3 NH3 has a remarkable feature: the conductivity as a function of increasing concentration has a gaussian like behavior with the presence of a maximum conductivity.

At the molecular level we showed theoretically, based on the experimental radial distribution functions (RDF), that the electrolyte could correspond to a mixture of both NaI molecule and dissociated NaI related to a dielectric constant of ammonia of 17.

3.2. Polyethyleneimine Modelization

Crystal structures of anhydrous and hydrated polyethylenimines offer a good starting model even some atomic coordinates are missing. Curiously, these polyalkylenimines are very used but few is known about physical-chemical properties. *Ab initio* calculations were made in this sense. Among the results, I checked they are large-gap insulating polymers, around 6 eV.

Concerning the electropolymerization mechanism, the first electrochemical stage of linear polyalkylenimine growth, undergoing oxidation mechanism via a radical cation is in good agreement with theoretical calculations. In the absence of water in the ethylenediamine based electrolyte, the system evolves towards a polymerization favorably with a $\Delta G < 0$ which is not the case in presence of water.

Patents

Herlem G., Gharbi T., Humbert P., «System for detecting at least one chemical substance» EP1535060 (2004).
Herlem G., Gharbi T., Humbert P., «System for detecting at least one chemical substance», US P.A.20060121598 (2003).
Herlem G., Gharbi T., Humbert P., «Bio-nez: système de détection d'au moins une substance chimique», FR2842604 (2002).
Herlem G., «Biosensor and method for making same», PCT/FR2001/001133. (2001).
Herlem G., «Biosensor and method for making same», WO/2001/077656. (2001).
Herlem G., Penneau J-F., Capitaine F., «Double layer high power capacitor comprising a liquid organic electrolyte», WO1998035369(1998).

References

Antoniou A., Herlem G., André C., Guillaume Y.-C., Gharbi T., "Simple method for detection", *Talanta,* 84, 3, 632-637 (2011).
Lakard B., Herlem G., Fahys B., «Electrochemical polymerization of 1,2-ethanedithiol as a new way to synthesize polyethylenedisulfide», *Polymer,* 49, 7, 1743–1747 (2008).
André C., Ibrahim F., Gharbi T., Herlem G., Guillaume Y.-C., "Experimental studies of OH", *Journal of Chromatography B,* 878, 28, 2826-2830 (2010).
André C., Herlem G., Gharbi T., Guillaume Y.-C., "A New Arginase Enzymatic Reactor: Development and Application for the Research of plant - derived Inhibitors", *Journal of Pharmaceutical and Biomedical Analysis,* 55,1, 48-53 (2011).
Picaud F., Herlem G., Girardet C., "Control of carbon", *J. Chem. Phys.,* 135, 154703 (2011).
Herlem G., Segut O., Antoniou A., Achilleos C., Dupont D., Blondeau-Patissier V., Gharbi T., «Electrodeposition and characterization of silane

thin films from 3-(aminopropyl) triethoxysilane», *Surface and Coatings Technology,* 202, 8, 1437–1442 (2008).

Herlem G., Zeggari R., Rauch J.-Y., Monney S., Torrealba Anzola F., Guillaume Y., André C., Gharbi T., "One-pot electrosynthesis of polyglycine-like thin film on platinum", *Talanta,* 82, 1, 417-421 (2010).

Herlem G., Monney S., Blondeau-Patissier V., Fahys B., Gharbi T., «Electrocatalyzed synthesis», *Electrochimica Acta*, 54, 27, 6797-6802 (2009).

Herlem G., Gharbi T., Ben Sedrine N., "Analyzing ab initio infrared spectra and electronic properties of polyethylenimine water", *J. Molecular Structure: Theochem,* 945, 1-3, 15, 27-32 (2010).

Boulanouar O., Khatyr A., Herlem G., Palmino F., Sanche L., Fromm M., "Soft Adsorption of Densely Packed Multi-Layer [DNA Plasmid – 1,3-diaminopropane] Complexes onto Hyghly Oriented Pyrolitic Graphite designed to Erode in Water", *J. Phys. Chem.,* B *115* (43), 21291 (2011).

Segut O., Herlem G., Lakard B., Blondeau-Patissier V., Nardin M., Gree S., Rauch J.-Y., "Electrochemically deposited polyethylenimine films", *Synthetic Metals,* 160, 11-12, 1359-1364 (2010).

In: Glycine
Editor: Wilhelm Vojak

ISBN: 978-1-62417-263-2

Chapter 6

CLINICAL USES OF GLYCINE

***Michael D. Davis*[1], *Brian K. Walsh*[2] *and John F. Hunt*[3]**
[1]Virginia Commonwealth University School of Nursing – Adult Health and Nursing System, Richmond, VA, US
[2]Children's Medical Center, Dallas, TX, US
[3]University of Virginia School of Medicine - Department of Pediatrics Charlottesville, VA, US

I. OVERVIEW

Glycine is a non-essential amino acid used commonly asa constituent of amino-acid based nutritional supplements (enteral formulas) in medical settings. As an example, standard infant nutritional formulas all contain glycine (1-2 g/100 g protein), in an effort to match breast milk - which contains 2.7 g glycine per 100 g protein (http://www.brightbeginnings.com/ -- Feb 2007), which amounts to approximately 35 to 40 mg per 5 floz feeding. Glycine is alsoused to lavage the genitourinary tract during gynecologic and urologic surgery. More recently, in the treatment of pulmonary hypertension, glycine has been used as an alkalinizingagent for inhaled prostacyclin, which requires alkaline conditions for stability. Presently, studies are evaluating the use of inhaled isotonic alkaline glycine buffer as a diluent for many inhaled therapeutic compounds and the use of intravenous glycine for amelioration of oxidative stress and ischemia/reperfusion injury.

II. Current Clinical Uses of Glycine

The most common clinical uses are as a diluent and a lavage. Glycine has distinct qualities that make it desirable for these uses. During endoscopic genitourinary surgical procedures, glycine is often used as a lavage fluid [1]. For intravenous and inhaled delivery of prostacyclin, glycine is used as diluent fluid [2, 3].

Glycine is used as a buffer, leveraging either its acidic or alkaline pKas. Since glycine is the smallest endogenous amino acid, it lacks an allergic potential. It also does not cause hemolysis and is transparent, two features desirable in a lavage fluid. For these reasons, glycine has been used to lavage during genitourinary surgical procedures since 1948 [4]. During these procedures, multiple liters of 1-2% glycine are often used to lavage the urinary tract and aid in endoscopic visualization [1].

Glycine is also used as a diluent fluid for both intravenous and inhaled delivery of the pulmonary vasodilator prostacyclin [2, 3]. Prostacyclin rapidly degrades in neutral or acidic environments, rendering it less useful when diluted in common pharmaceutical diluents. Glycine has an alkaline pKa value of 9.78, allowing it to dilute prostacyclin while maintaining a moderately basic pH [5].

When discussing clinical uses of glycine, as with any medication, it is important to note potential side effects. Micromedex describes a phenomenon referred to as "prostatectomy syndrome." This syndrome is related to the use of large volumes (1-3 L) of usually 1.5 – 2.2 % glycine solution. Extravascular and/or intravascular extravasation of this solution can lead to cerebral edema, bradycardia, chest pain, electrolyte abnormalities, and temporary blindness. The data from the glycine used in urologic surgery has also produced the only pharmacokinetic data describing glycine, although it is of course not necessarily relevant to other routes of administration. Depending on the volume given, the half-life of intravenous glycine appears to be between 85 and 100 minutes. The longer half-life was associated with the larger volumes and higher concentrations [6].

Elimination of glycine begins with its catabolism to methylene tetrahydrofolate, carbon dioxide, and ammonia. The production of ammonia has alerted clinicians to use caution when administering glycine to patients with liver or renal insufficiency. This concern is non-existent when delivering glycine via an inhaled route due to the very low total quantity of glycine delivered in this fashion compared to urologic, intravenous, and enteral uses of glycine [7, 8].

III. Research Applications of Glycine

Due to its alkaline pKa value, interest has arisen in use of glycine as a diluent for inhaled medications other than prostacyclin. Airway acidification is associated with a variety of pulmonary disorders and disease processes [9]. Also, many inhaled medications, including beta-agonists and anticholinergics, transport across airway epithelium more effectively in an alkaline environment [10]. Knowledge of the role of airway pH in pulmonary health, along with the development of devices and techniques to measure it, has created interest in treatment of airway pH disturbances. Improved ability to treat or potentially reverse acidic airway pathology by means of therapeutic alteration of airway pH could have an impact in respiratory medicine. Specific Phase 1 and 2a studies of inhaled alkaline glycine have demonstrated safety and therapeutic potential; clinical trials are underway evaluating the use of inhaled alkaline glycine in conjunction with aerosolized albuterol for the treatment of obstructive lung disease.

Dietary supplements containing glycine are being evaluated for their effects on oxidative stress and also on uncontrolled type-II diabetes. These supplements have been shown to decrease oxidative stress and aid in diabetes control [11, 12]. Studies are also evaluating the use of intravenous glycine on liver damage. These studies indicate that glycine may have benefit in reducing organ damage during transplant and after ischemic events [13, 14].

References

[1] Hahn RG. Irrigating fluids in endoscopic surgery. *Br. J. Urol.* 1997;79(5):669-680.

[2] Olschewski H, Simonneau G, Galie N, et al. Inhaled iloprost for severe pulmonary hypertension. *N. Engl. J. Med.* 2002;347(5):322-329.

[3] Siobal M. Aerosolized prostacyclins. *Respir Care* 2004;49(6):640-652.

[4] Rhymer JC, Bell TJ, Perry KC, et al. Hyponatraemia following transurethral resection of the prostate. *Br. J. Urol.* 1985;57(4):450-452.

[5] Sigma-Aldrich. *Glycine - Material Safety Data Sheet.* 04-24-2012;410225.

[6] Norlen H, Allgen LG, Vinnars E, et al. Glycine solution as an irrigating agent during transurethral prostatic resection. Glycine concentrations in blood plasma. *Scand. J. Urol. Nephrol.* 1986;20(1):19-26.

[7] Habler O, Kleen M, Takenaka S, et al. Eight hours' inhalation of prostacyclin (PGI2) in healthy lambs: effects on tracheal, bronchial, and alveolar morphology. *Intensive Care Med.* 1996;22(11):1232-1238.

[8] Habler O, Kleen M, Zwissler B, et al. Inhalation of prostacyclin (PGI2) for 8 hours does not produce signs of acute pulmonary toxicity in healthy lambs. *Intensive Care Med.* 1996;22(5):426-433.

[9] Ricciardolo FL, Gaston B, Hunt J. Acid stress in the pathology of asthma. *J. Allergy Clin. Immunol.* 2004;113(4):610-619.

[10] Horvath G, Schmid N, Fragoso MA, et al. Epithelial organic cation transporters ensure pH-dependent drug absorption in the airway. *Am. J. Respir. Cell Mol. Biol.* 2007;36(1):53-60.

[11] Sekhar RV, Patel SG, Guthikonda AP, et al. Deficient synthesis of glutathione underlies oxidative stress in aging and can be corrected by dietary cysteine and glycine supplementation. *Am. J. Clin. Nutr.* 2011;94(3):847-853.

[12] Sekhar RV, McKay SV, Patel SG, et al. Glutathione synthesis is diminished in patients with uncontrolled diabetes and restored by dietary supplementation with cysteine and glycine. *Diabetes Care* 2011;34(1):162-167.

[13] Wang YS, Yan YH, Zou XF. Protective effect of glycine on liver injury during liver transplantation. *Chin. Med. J.* (Engl) 2010;123(14):1931-1938.

[14] Petrat F, Boengler K, Schulz R, et al. Glycine, a simple physiological compound protecting by yet puzzling mechanism(s) against ischaemia-reperfusion injury: current knowledge. *Br. J. Pharmacol.* 2012; 165(7):2059-2072.

INDEX

E

F

G

H

I

K

L

M

Q

R

S

T

U

V

W

X

Y

Z